Life on a $5 Bet

★★

Major General

Edward J. Mechenbier

USAF (Ret.)

LIFE ON A $5 BET

★★

MAJOR GENERAL

EDWARD J. MECHENBIER

USAF (RET.)

Edward J. Mechenbier
Major General, USAF (Ret.)

with
Linda D. Swink

LITTLE MIAMI PUBLISHING CO.
Milford, Ohio
2012

Little Miami Publishing Co.
P.O. Box 588
Milford, Ohio 45150-0588
www.littlemiamibooks.com

Copies of this book can be obtained by contacting the publisher.

Printed in the United States of America on acid-free paper.

978-1-932250-98-5 Hardbound (2012)
Library of Congress Control Number: 2012937973 Hardbound (2012)

ISBN 978-1-941083-22-2 Paperback (2019)
Library of Congress Control Number: 2019953274 Paperback (2019)

Dedication

This book is dedicated to the spirit, strength, and sacrifice of those who died to make it possible for us to live the life we do, and to all those who serve today in the military and public sector preserving those gifts.

I salute you.

Contents

Foreword

ON 30 JUNE 2004, I recall Major General Ed Mechenbier, USAFR, telling an audience of more than a thousand professional associates, ardent admirers, friends and family members: "When you're being run out of town on a rail, it's always best to be out in front and act like you are leading the parade!"

The occasion was General Mechenbier's formal retirement from the United States Air Force Reserves, conducted at the National Museum of the U.S. Air Force in Dayton, Ohio. I was the presiding official and during my remarks, I recall mentioning that "Were it not for the statutes governing the service of Reserve Officers, which required they retire at age sixty-two, Ed Mechenbier would continue to serve his nation in uniform until he no longer had the strength to perform his duties."

I also remember commenting that from the moment I first met Ed and learned that he had been a Vietnamese prisoner of war (POW), it was my observation that Ed and his most extraordinary wife, Jerri, may have experienced a *defining moment* when Ed was shot down and incarcerated for nearly six years. But both he and Jerri were not *defined* by those experiences. No, in fact, as you will read in this sobering, exhilarating and extremely meaningful biography, Ed and Jerri Mechenbier are so much more than can be described by their 1967 to 1973 experiences . . . and you will be uplifted by how they have knitted those experiences together with other *defining moments* in their lives to become one of this nation's most devoted and cherished couples.

As you will learn, Ed went from a family of eight children in Ohio, with hardly an extra dollar to be had, to the United States Air Force Academy as a result of a $5 bet with his dad. At the time, I'm not sure Ed knew how much confidence his dad had in him, but as a result of that confidence, and a $5 bet, Ed's transition from a Midwestern boy to young officer in the Air Force was going to be a significant *defining moment* in his life. From the deliberate actions of the upperclassmen bent on transforming each new cadet's penchant for being insular and self-reliant when under stress, to an individual who learns to depend on the team and cooperation to overcome adversity, you will gain an insight on how Ed learned that the Air Force Academy's officer corps bent over backwards to ensure his success in completing the Academy's rigorous academic, military, and physical training regimen. Ed was being molded into an officer and leader for life.

Equally *defining* for Ed were his introduction to Jerri not long before being accepted to the Academy as well as his first flight in an Air Force jet aircraft shortly after arriving at the Academy. In both cases, he found a love and zest for life that inspired, energized, and sustained him during some of the most exciting days and during some of the darkest days of his life. These two events framed Ed's sense of purpose and devotion for life.

Of course, as you read this fascinating biography, you will learn—and feel—the fear, the pain, the humility and the sense of despair and hopelessness faced by every one of our brave military members who have been incarcerated during combat operations. To say that circumstance is a *defining moment*—for both those POWs and their families—is an understatement. As with many other POWs, you will find as you read this book, neither Ed nor Jerri allowed those years of questioning, doubt, pain, hope, and ultimate joy define them. Although in order to survive those years, they had to demonstrate incredible courage, discipline, trust, and patriotism. They never let those years tether them to a field of self-pity or entitlement.

No, instead they used those experiences to grow their circle of friends and influence while advancing to greater levels of leadership, responsibility, and command in the United States Air Force, the Ohio Air National Guard and finally in the Air Force Reserves. In fact, it was my experience that Ed and Jerri offered many *defining moments* for literally thousands of other military and civilian members of our military services.

I know that, because when I became the Commander of the Air Force Materiel Command in 2003, Ed Mechenbier was my mobilization assistant; an Air Force Reserve major general assigned the duty to not only help advise me on the best use of our several thousand Air Force Reservists who served the Air Force Materiel Command, but also to replace me during periods of increased tension, to include war, should I and my vice commander be asked to assume larger responsibilities as more and more of our combat units and senior leaders deployed forward to a war zone.

During our first meeting he gave me some great advice when he said, "General, you'll find this command to be a little different than your last one in Europe as combat commander. The Air Force and the DoD [Department of Defense] think of this command as a business that really needs to be managed. Although managerial duties are important here, what this command really needs is a good leader and I'll do everything I can to make sure you have time to be a good leader." That was a *defining moment* for me. And the day before he retired, at the end of our last meeting, he shook my hand and said, "Keep Leading!" I'm not sure I have ever felt more proud than to have been complimented by Ed Mechenbier.

I know you will enjoy this great biography and I hope you'll find some *defining moments* to inspire you in the years ahead.

Gregory S. Martin
General, USAF (retired)

Introduction

THE WORD "HISTORY" CONJURES UP IMAGES of dusty events of long ago, the value of which is usually considered more entertainment than enlightenment. We have textbooks and lecture series to educate and explain what we must know. The Vietnam War is now approaching four decades since its conclusion in 1975, so it's just about ready to qualify for that "old and barely interesting" category. Of course, there are friends and family who will read this book and perhaps learn a little about Ed Mechenbier they didn't know or even suspect. But what is inside these pages that might hold the interest or be worthy of the time of other readers? Certainly as an individual, I am not the stuff of whom epic stories are told or someone whose individual contributions have made a measurable impact on the course of events.

This biography is not so much about presenting my life story but to shine a light on the character and events of the men who served in the POW camps of North Vietnam. In a grander sense, it also provides some insight into the type of men and women who serve in our military today.

Life on a $5 Bet had its genesis from the book I had wanted to write titled *Ho Ho Ho Chi Minh* when I first came home from being held captive by the North Vietnamese. Like so many other returning POWs who wrote about their experiences as a POW, I, too, wanted to share that experience, not to belabor the horrendous and torturous conditions we lived under, but how, by holding to the Code of Conduct and a spirit of brotherhood, we endured.

Many of those books written by my fellow prison mates told of the long, miserable days at the hands of cruel and hateful guards and camp commanders who gave no thought to pulling our arms up behind and over our heads until our shoulders popped from their sockets. For the most part they recount the same experiences I endured. They are true and accurate accounts. Although these events happened, I wanted to tell another side of prison life—the funny side. Yes, funny things did happen to us, and humor became a mechanism that made the serious aspects of prison life more palpable.

I hope the reader will see the role humor played in our survival. Whether the torture experience filled 10 percent of the time or 90 percent of the time, the ability to "bounce back" as told in Geoff Norman's excellent book,[1] humor recharged our spirits and our bodies for the next physical and emotional challenge.

Seeing the funny side of prison life when I was being tortured was tough, but when the ropes eased and the beatings stopped, the ability to keep a perspective, to see the positive, and, yes, to see the humor that got me through the bad parts went miles in steeling me to survive. It all boiled down to looking at each situation, and trying to see the irony in what all of us were experiencing.

There's an old saying, "There are no extraordinary men just ordinary people who rise to extraordinary situations." I was put into a horrible situation and I found a way to adapt, to cope, and to win in exactly the same manner as the other 591 POWs.

The reader of this book should know that Ed Mechenbier is not extraordinary. I'm no hero, I just happen to lose a five dollar bet and came within two seconds of dying. Now I look at the rest of my life as a gift. I still have six inches between my ears, sometimes full of putty, sometimes filled with a pretty good carbon-based computer. I have physical, emotional, and mental strengths and weaknesses like everybody else.

This book is presented not as a testimonial to one man's life or even as a saga of the life and times of the subject; it was undertaken to reveal the strength of character innately resident in all of us. To be sure, there are passages that may sound beyond belief, but they were real and tolerable by not just me but those with whom I served.

From my experiences as a POW I continued to serve in the United States Air Force, Ohio Air National Guard, and the Air Force Reserve, retiring as a major general. Not too shabby for a kid who had to borrow money from his mother to pay a five dollar bet he made with his father. But that bet started me down a career path even the best fiction writers could not have imagined. For that, I am eternally grateful. After all, I am still the world's greatest fighter pilot.

The events of my imprisonment occurred more than forty-four years ago. My memory of those events has clouded somewhat over time, but I have recounted them as best I could along with what I'm sure are a few embellishments. The more recent events are clearer, although there may be others who might questions my memory. I have changed the names of several characters to protect their identity and my hide.

---

1. Geoffrey Norman, *Bouncing Back: How a Heroic Band of POWs Survived Vietnam* (Boston: Houghton Mifflin, 1990).

CHAPTER 1

Bail Out!

June 14, 1967

I HAD JUST DROPPED TWO TONS OF BOMBS on the Vu Chu railroad yard in North Vietnam when Kevin McManus, my backseater, keyed the mike and said, “Looks like we have a half dozen SAMS (Surface to Air Missiles) on our radar.”

We needed to get the hell out of there.

Before rolling in on the target, I had set one engine at idle, and kept an eye on the airspeed indicator to know when to light the second afterburner. My plan was to make a sharp turn, head out to sea, and then link up with a KC-135 to refuel before returning to Da Nang.

When the airspeed reached 600 knots, I lit the afterburner, but instead of acceleration, I heard, BAM!

That’s when all hell broke loose.

Huge shrapnel-filled puffs of black smoke exploded around my F-4 Phantom as antiaircraft fire pounded the air and 37mm shells streaked past, just missing the canopy. Instinctively, I ducked from side to side. I suddenly found myself in the old World War II movie, *Twelve O’Clock High*, piloting a B-17 and being shot at after bombing cities in Europe.

The red firelight on the control panel began flashing, telling me I had a fire in the right engine. Glancing in the rearview mirror, I saw what no fighter pilot ever wants to see—a raging inferno chasing me. It was my own airplane.

I pushed the red button to extinguish the fire, but nothing happened.

When I flipped the engine’s master switch to shut down the engine, the airplane suddenly made a snap roll to the right, throwing me hard against the inside of the cockpit.

To correct the roll, I swung the stick to the left but it just flopped uselessly back and forth. The ailerons were gone.

With the airplane upside down and losing altitude, I tried the elevators. Nothing. The hydraulic system was shot.

Not about to give up or let fear take over and drive me to do something stupid that would fry my ass, discovering the rudders still worked, I stomped the left rudder. The airplane stopped rolling and leveled out, but then started to ratchet as it rolled, stopped, and rolled again.

When the airplane spun the second time, I pounded the rudder again. The damaged airplane steadied, but only for a second before it began corkscrewing through the air a third, fourth, and fifth time. I struggled to regain control, but the aircraft continued spiraling and gaining airspeed as it headed for the ground.

Below 6,000 feet, we were in danger of getting hit by small arms fire and with all the aircraft's systems shot, I was out of options, left with only my skill and training to pull us out of this mess.

At 4,500 feet, the airplane rolled the sixth time. That's when Kevin called through the headset, "Ed, I don't think we're going to make it."

Still fighting the controls, I wanted to yell at him to shut up and leave me alone. Everything will be okay. I'm a fighter pilot for God's sake. There ain't nothing I can't fix. I'm in control here. I can handle this.

"Ed, I don't think we're going to make it." This time Kevin's voice was adamant.

Kevin's urgent nudge brought me to my senses and at 4,000 feet I gave the order, "Bailout, Kevin! Bailout!"

Kevin didn't wait for the second bailout order. "Sayonara, Ed," he said, and pulled his ejection seat handle at the same time I pulled mine. Our seats shot out of the airplane simultaneously, but upside down. Not a textbook ejection.

A blast of air hit me as I rocketed from the airplane. Leaving an F-4 with a Roman candle attached to my rear end was like being shot out of a cannon with twenty-one men sitting on my chest. The pressure of the g-force squeezed so hard I thought my lungs would collapse.

It took only one and three-quarter seconds from the time I ejected until my parachute snapped open. And before I had a full chute, my airplane hit the ground, exploding into an orange ball of flame. Without Kevin's warning, I would have ridden the airplane into the ground without ever giving a second thought to bailing out.

Hanging in my parachute harness, I looked up and saw holes in the canopy. Looking down I found out why. Angry peasants were taking potshots at me. I was a clay pigeon on a skeet shooting range floating right into their midst. All I could do was pray please don't shoot me.

My heart pounded as I drifted toward a small farming village surrounded by a patchwork of rice paddies. In any other situation, the scene below would have been a photographer's delight, a pastoral postcard, but with bullets sailing past my head, the picture was anything but serene.

I disconnected my facemask, pulled my radios from my survival vest, and then smashed them together, breaking the batteries and antenna so the Vietnamese couldn't use them to lure rescue helicopters into a trap. I detached the survival kit from behind my legs and watched my food, water, raft, poncho, and signaling mirror, all the equipment needed for survival, fall below me.

Having a .38 caliber pistol as my only protection, and not wanting those peasants to think I was going to shoot it out with them when I hit the ground, I threw the pistol away. It probably landed in a rice paddy somewhere and may still be there for all I know.

With my parachute full of holes and no way to maneuver it, I was at the mercy of the wind. As my luck would have it, I landed on the pitched roof of the only building within sight. I rolled off, falling twelve feet and tangling myself in the shroud lines on the way down. I lay on the ground like a Christmas package wrapped in a silk cocoon. So much for all my training hanging for hours in a harness practicing the correct parachute landing—eyes on the horizon, feet together, and knees slightly bent. Nice in theory; not practical in reality.

I saw Kevin make a perfect two point landing in an open field not far away, but before he had gathered his chute, and before I could free myself from my tangled web, we were surrounded by a group of nonuniformed militia carrying every conceivable type of weapon from high-powered automatic rifles to guns so old they could have been turn-of-the-century flintlocks, and they were all pointed at us.

From the menacing expressions of these rebels, it was apparent they weren't going to invite us to dinner. Article II of the Code of Conduct says: *I will never surrender of my own free will*, but with a hundred guns aimed at me, it was hard to muster my free will.

Two men wearing belted pants and buttoned khaki shirts yanked off my helmet and parachute harness, and then, with what seemed like thirty-foot long machetes, started cutting through the shroud lines. Once I was untangled, one man pulled me to my feet and began slashing through my g-suit. He then ripped off my flight suit as another man stripped me of every stitch of clothing except my tee shirt and shorts. They didn't stop there; they took my dog tags, Geneva card, boots, socks, gloves, and wristwatch.

Hundreds of peasants emerged from the rice fields and surrounded us to watch the proceedings. Several barefoot children, too young to appreciate the political aspects of war, threw rocks at us, while others gathered around, staring at the two strange men who had fallen from the sky. A group of middle-aged citizens wearing black pajama-like clothes and conical hats yelled at us, and although I didn't understand what they were saying, their angry shouts and fist waving suggested they were a bit unhappy that we had just dropped in. Interestingly, however, a circle of older people stood off to one side, arms folded, with no expression of animosity as if to say, "Sorry about your luck, fellows."

I stood bewildered as someone in the group began snapping photographs; I saw no reason to smile. Once satisfied he had enough pictures, he turned and melted into the crowd.

Then a strange thing happened. From behind me a man's voice whispered in my ear in perfect English, "What is your name?"

"First Lieutenant Edward Mechenbier," I answered.

That was all he said. I never heard that voice again, and there was never another attempt to make contact with me.

A man carrying a handgun came forward and said something to me in Vietnamese. I had no idea what he was saying and just stood looking at him. He pointed at my shorts and undershirt and then motioned with his gun. When I didn't move, he began tugging at my shorts. I got the message and removed my underwear and then stood there naked as a jaybird in the afternoon sun. I was given a strip of blue cloth to tie around my waist, while a man hogtied my hands behind my back with a coarse rope that dug into my wrists.

The third article of the Code of Conduct says: *If I am captured I will continue to resist by all means available.* This was where I was supposed to come up with some clever method to avoit being captured, but standing there wearing only a loin cloth, I saw no means available.

Kevin and I were separated. I was taken to a small thatched hut that looked like someone's home and pushed inside. Bamboo mats hung from the low ceiling. A stool, a small square table, and a chair were the only furniture in the room. A windup clock sitting on the table read 2:15 p.m. It had been only fifteen minutes since I had dropped my bombs on the railroad yard.

Villagers peered through the tiny window, some made faces, while others just stared at me as I sat wondering what was going to happen next. Nothing did, so I waited.

It seemed forever before the door opened and two Kahunas came in, pulled me to my feet, and with guns drawn, ushered me outside.

I saw Kevin being led from another hut nearby.

"Hey, Kevin. You okay?" I called.

Before Kevin could answer, the barrel of a rifle was stuck in my back, making it clear that talking was forbidden. Kevin gave me a halfhearted grin to let me know he was all right.

The third article of the Code of Conduct also states that *I will make every effort to escape and aid others to escape.* I looked over at Kevin, wondering if he had a plan because I certainly didn't. After the intense, weeklong jungle survival training received in the Philippines, I should have been able to escape, but with a gun in my ribs and my hands tied, escaping seemed futile. There went that theory.

With secured ropes around our wrists, and satisfied we couldn't escape, they began force-marching us over a dirt road. We were herded like cattle for what seemed miles as the sun scorched my head, back, and shoulders. The heat seared my lungs making breathing difficult.

After several hours with no rest, my back, injured during the ejection, began to hurt like hell and my legs felt like anchors. My bare feet, unaccustomed to running over jagged rocks and sharp sticks were cut and bleeding.

Feeling faint, I stumbled and went down on one knee. My captor, taking advantage of the situation, kicked me in the ribs. I doubled over only to be kicked again.

Kevin bravely limped along beside me sweating and panting, obviously in pain, too, but our discomfort seemed of no concern to our captors as they spurred us along by sticking sharp pieces of bamboo into the backs of our legs. When we could no longer walk, they dragged us.

Exhausted, we finally reached a hamlet on the edge of a mountain where a hostile welcoming committee came out of their hooches and greeted us with pitchforks, rocks, and mud balls. Some pelted us with stones. Others spit on us while our captors stood by proudly showing off their two American trophies.

Bruised and bloody, Kevin and I were led to two small, shallow trenches that looked like bomb shelters and, in broken English, told that these were our graves. The thought struck me, how were they going to fit my six-foot frame into that tiny hole?

Three men carrying rifles lined up forming a firing squad and stood ready to take aim.

"Now, you die," the leader said.

I was too tired and frightened to do anything but resign myself to the fact that I was about to be shot.

The leader motioned for people to stand back and began counting, "Một. Hai," one, two in Vietnamese, and then called, "Ready aim."

Looking out over the valley I saw a beautiful statue of the Blessed Virgin silhouetted against the sky. She looked sad. I said a Hail Mary, praying she would intervene in my hour of need, and then closed my eyes and braced myself for the report of the rifle.

But nothing happened.

When I looked up, curious on lookers were moving closer taking pictures, having a Kodak moment.

The man in charge gestured again for everyone to stand back and be quiet. "Okay, now you die."

I readied myself again.

The firing squad raised their guns.

"Một. Hai," he said, but then paused and yelled "Stop! More pictures. More pictures."

The crowd pressed in for another round of picture snapping.

Satisfied at last that everyone had taken enough photographs, he said, "Now you die. Một. Hai,"

I waited, hoping the bullet wouldn't hurt when it entered my body.

Again, the leader paused and waved the children out of the way.

It was taking a long time to die.

When everyone started laughing, I realized the scene was just a set up, a mock execution orchestrated for propaganda. After returning my underwear—why they took them in the first place was beyond me unless it was to humiliate me—they forced Kevin and me to the side of a building and shoved us against the wall where we became a carnival sideshow, the entertainment for the day.

Come one. Come all. See the funny looking Americans.

We stood helpless as small children used us for target practice throwing live rats and mice at us. Adults shouted as they jostled forward to stick lit cigarettes into our arms and legs.

I don't know how long we stood pressed against that building with rats slithering at our feet, but night had settled over the village and the sky had darkened to a deep purplish hue by the time uniformed soldiers from the regular army appeared and rescued us. I say "rescued" because I believed they were saving us from these terrorizing peasants. I was actually relieved to see the soldiers. Up to that point I was never sure what this ragtag bunch of civilians were capable of. At least with the regular army we could expect a semblance of military structure. We would officially be prisoners of war and regarded as such under the rules of the Geneva Convention.

With our hands still bound behind our backs, soldiers stood guard with automatic weapons while our feet were tied with heavy rope. We were blindfolded, put into the back of a truck, and then taken on a bumpy ride to what I presumed to be Kep Airfield not far from the railroad yard we had bombed earlier that day.

At the field, I heard rotor blades of a Soviet helicopter beating the air as they rolled us head first and upside down into the cabin. Riding in a helicopter tied and blindfolded was bad enough, but upside down in a swaying vehicle was like riding a roller coaster and a perfect way to lose your breakfast, which I did.

After a short flight we were taken from the helicopter and thrown into the back of a small vehicle where Kevin and I jostled along—squeezed together like two lovers in the back seat of an old Chevy. Sounds of the road underneath the tires changed from a rumble to a steady thrum, leading me to believe we were traveling into a city. I could make out the swoosh, swoosh sound of steel girders as we crossed a bridge.

Once across, tiny bells, like those on a kid's bicycle blended with vehicle horns and the chatter of people speaking in Vietnamese. This unusual blend of sounds seemed strange in the middle of the night.

With an assignment to Vietnam, and wanting to know about the country, I had read several books by Bernard B. Fall who described the culture, tradition, history, and political posture of Vietnam, and felt prepared for what to expect. What Fall failed to describe was the putrid smell. A very distinct stench between that of open sewage, rotten cabbage, and five-day-old fish, permeated the air.

Unexpectedly, the driver hit the brakes and the vehicle came to a sudden halt. With my hands tied, and with no way to hold on, I was thrown forward and rebounded, almost tumbling out.

Then I heard the sound of a tiny horn and muffled voices. Metal scraped against metal and the vehicle began moving again. When it stopped the second time, Kevin and I were pulled out. A soldier stuck a rifle in my back and pushed me forward, but with my feet bound, all I could do was shuffle a few steps at a time.

Kevin mumbled under his breath, but then his voice trailed off and I sensed he was no longer near me. I wanted to call to him, but reconsidered.

I had taken only a few steps when I heard the rattle of keys being inserted into a lock. A door squeaked opened and I was pushed inside. My hands and feet were untied and the blindfold removed. The door slammed shut. I looked around.

I was alone.

CHAPTER 2

Hello Hanoi

HOA LO PRISON, KNOWN AS THE HANOI HILTON by American captives held there since 1964 when naval aviator Lieutenant (jg) Everett Alvarez, Jr., was shot down, was built by the French between 1886 and 1898 and expanded as needed during their rule. The compound, in the heart of downtown Hanoi, was made up of several sections each having a name given by the Americans such as Heartbreak Hotel, New Guy Village, Camp Unity, and Little Vegas, which was further broken down into the Stardust, Mint, Desert Inn, Thunderbird, Riviera, Golden Nugget, Courtyard, and the Tet Room.

A gloomy place at best, surrounded by a fifteen-foot-high concrete wall, Hoa Lo was the first prison most prisoners were taken for screening, interrogation, and indoctrination into the political mindset of the North Vietnamese who considered the Americans "war criminals."

Atrocities that included beatings, physical and psychological torture, starvation, and neglect were common. Men often returned from an interrogation session with broken bones, missing teeth, shattered eardrums, and dislocated limbs. Some men didn't return at all. To this day the Vietnamese deny that torture ever occurred there.

I found myself in a small, filthy cell with no bed and no place to sit. A bare, dim light bulb hung from the ceiling, making it hard to see what might be lurking in the dark corners. With no windows or obvious source of ventilation, the place stunk. Stagnant air coupled with the oppressive heat and humidity made me sick to my stomach.

Somewhere close by I heard a heavy thud followed by a series of thumps and bangs, and then sounds of flesh against flesh. Someone screamed and then moaned. Listening closer, it sounded like Kevin. What in the name of God were they doing to him? I felt for him, but there was nothing I could do. When the moaning faded, I waited for a sound, anything, that would tell me he was okay, but all I heard were distant voices talking in Vietnamese. I prayed they hadn't killed him.

I spent the rest of the night sitting on the tiled floor leaning against the wall of that hot, foul-smelling cell with my back killing me. A nice hot, soaking bath would have felt good but I doubted that luxury was in my future anytime soon. At least my feet, although covered with dirt, had stopped bleeding.

I don't remember falling asleep, but the sound of door opening woke me. A turnkey, dressed in a faded green shirt several sizes larger than his thin frame, pants cinched at the waist by a cord for a belt, and rubber sandals, stepped into my cell and set a bowl on the floor in front of me, and then pointed to it, indicating I should eat. My first meal, a gourmet's delight was lukewarm cabbage soup cooked in what looked like brown ditch water with something floating on top. Looking closer, I realized it wasn't croutons but bugs, and they were staring up at me. My last meal before setting out on my mission to bomb the Vu Chu railroad yard, now many hours ago, had been ham, potatoes, and green beans. My stomach rebelled looking at the slop they expected me to eat, but I was hungry, and as foul as the food was, I ate it, a la bugs and all.

The guard watched until I finished eating, and then offered me a cigarette. I didn't smoke, and wasn't about to take anything the Vietnamese offered. Taking the cigarette could be construed as accepting a special favor and in violation of the third article of the Code of Conduct: *I will accept neither parole nor special favors from the enemy.*

I heard nothing the rest of the day or that night. A day and a half went by, still nothing. The only jailer I saw was one who brought me food and a little porcelain bucket like the ones you see in old western movies to use as a latrine.

Since no one was bothering me and I was not being mistreated, I decided maybe this wasn't so bad and began fantasizing that I would soon be released. Perhaps they were trying to figure out what to do with Kevin and me and were looking for a way to turn us loose.

Yep, that's what I thought, so I waited.

The Interrogation Begins

The next day, my hopes were dashed when a guard entered my cell, yanked me to my feet, stuck the barrel of his rifle in my side, and then pushed me toward the door. I was taken down a grimy hallway to a small, windowless room, and made to sit on a small stool in front of a cloth-covered table while guards squatted close by holding guns with fixed bayonets.

A man who I assumed was a camp official sat at the table in a higher, superior position. He wore no markings on his clothes to indicate rank. He looked at me over a blue book lying on the table, taking his time trying to appear important—the dramatic pause—before finally asking in broken English, "What your mission?" He sat holding a pen ready to write my response.

I gave none.

"What airplane you fly?"

I looked straight ahead.

"Why you no answer?" he asked.

"I am an American fighting man and under the Geneva Convention, Article V, as a prisoner of war I am only required to give you my name, rank, service number, and date of birth."

"We not subject to Geneva Convention. You not prisoner of war. There is no war. No diplomatic relations between your country and my country. You are criminal. You bomb our women and children. You will be tried as criminal. Your fate in my hands now."

"Say again, what airplane you fly?"

"Edward J. Mechenbier, first lieutenant, 78807A, June 29, 1942."

"What is speed of aircraft when you bomb us?"

"Edward J. Mechenbier, first lieutenant, 78807A, June 29, 1942."

"How accurate your weapons?"

"Edward J. Mechenbier, first lieutenant, 78807A, June 29, 1942."

"You not understand. We in control here. You must answer question."

I made no comment.

He persisted. "Why you bomb peace-loving people of Vietnam? Don't you understand us? What is target for tomorrow?" He raised his voice an octave.

Of course, I didn't know the target for the next day. I doubted even the president knew from day to day.

He went on relentlessly with, "Tell me. Tell me. Tell me." When I didn't answer, he said, "You no answer. That okay. We know all answers. There is nothing we don't know." He seemed pleased with his pronouncement.

When I continued refusing to answer, his face contorted into a sneer and he yelled, "You know we can kill you."

This guy was an amateur; I had been yelled at by experts. My first day at the Air Force Academy I was met by an upperclassman who got his message across that he was in charge by hollering at the newly arrived recruits, telling us that we were the lowest form of life, unworthy of existing.

The shouting lasted all eight weeks of basic training as seasoned upperclassmen placed us in an intense physical and emotional environment to let us know that we had stepped into the far side, and yelling was just part of the game.

Now, here was this short, little weasel trying to break a tough Air Force Academy graduate. It was all I could do to keep from laughing as he shrieked at me in his butchered English.

The interrogation continued as he attempted to pry information from me that the Vietnamese could use for propaganda. They wanted this "Yankee Air Pirate" to malign our leaders and denounce capitalism. They wanted me to sing the praises of Communism and accept Ho Chi Minh as my savior. Yeah, like I would ever do that.

This was going to be a piece of cake. Using a technique learned at the Academy, I made a game of it by picturing him doing something stupid in his underwear, or by telling myself he looked like a funny animal or someone making a speech with spinach stuck between his teeth. Those visions boosted my confidence and reserve as I held my stance. I wasn't going to allow him to exploit me for his cause.

"Edward J. Mechenbier, first lieutenant, 78807A, June 29, 1942."

"You being foolish. You alone here. Nobody know you here." He stood, leaned over the table, looked down at me and said, "You in our hands now. You will be punished."

With that he turned and walked out.

Knobby Torture Room

True to his word, several guards hauled me out of the room and took me across the hall to a cell covered from floor to ceiling with greenish-blue, tennis-ball size blobs. I noticed a pile of rusted iron manacles lying in the corner and what looked like meat hooks hanging from the ceiling.

I quickly realized we were not going to have an intellectual discussion. Ignoring the Geneva Convention, the guards wasted no time employing a technique of physical persuasion to extract information from me called "beating the crap out of me."

They didn't ask a question and then wait for an answer; they just began using me as a piñata, clubbing me with their rifle butts. Each blow sent a sharp pain down my already injured back.

"You give answer, now?"

"No."

That must have been the wrong answer. One of the guards backhanded me across my face. Another kicked me in the stomach, bringing me to my knees while another pulled my head up by my hair and placed a swift kick to my jaw, cracking one of my teeth. I spit out the piece of the broken tooth, along with blood, onto the floor. Although the cell looked as though it hadn't had a good cleaning in years, spitting blood on the floor apparently was not to be tolerated. They increased the intensity of punching and kicking. With my body absorbing each blow, the pain in my mouth was the least of my problems.

"Answer question, now?"

I managed a resolute, "No."

That brought another round of punching.

When the guards finally stopped thrashing me, the officer came in, and with no emotion, said, "Now, you understand we have power to kill you. You will answer questions." And, like before, he turned and left the room.

Guards stood watch over me for the next three days, and each day that same officer came to ask if I was ready to talk.

When I continued to resist, they upped the ante and brought out the ropes to establish the rules of the game. Several musclemen appeared from the back of the room. One threw me onto the floor, another tied a rope around my ankles while a third guard pulled my hands behind my back, bound them together, and then lashed the other end of the rope tightly around my elbows. In a matter of seconds I was immobilized.

They stuck a bamboo pole through the knot in the rope and began pulling my arms over my head toward my feet until my face touched the floor. A searing pain tore through my arms and down my back as my left shoulder popped from its socket. I screamed, wishing I could die to release myself from the agony.

"Now, criminal will answer question," the sadist said.

With all the breath I had left I said, "You can kill me today, but I'm not going to tell you anything."

Vietnamese Rope Torture.
DRAWING BY MIKE MCGRATH,
COURTESY OF THE UNITED STATES NAVAL INSTITUTE.

Thunderbird 7

I thought they got the message that I wasn't going to cooperate and play their game, but apparently they failed to understand that Americans don't give in easily, if ever. Two guards dragged me out of the cell and into a dirt-covered courtyard.

Squinting into the sunlight, I saw for the first time the place I was being held captive. There was no doubt it was a prison. Thick brick walls with ass-slashing glass shards lined the top, sharp enough to do serious injury should anyone try to climb over to freedom.

They escorted me past a large, iron gate and into another part of the prison, and then into a building with high, boarded windows. I heard no sounds as they led me down a dark passageway with walls blackened with what looked like tar. The guards stopped in front of a door covered with peeling green paint, opened it, and shoved me inside.

The cell was similar to the first one in which I had been held—small, dirty, and stunk of fear and despair. Walls that showed evidence of once having been whitewashed were now tarnished with years of grime and neglect.

The guards threw me onto a concrete slab attached to the wall. One untied the rope around my wrists while another secured them behind my back in tiny, iron manacles. After bolting my feet into metal foot locks they walked out, slamming the door behind them.

I spent that first night lying on my back assessing my situation and came to the conclusion that my status had dramatically changed. I was no longer the invincible white knight riding my charging steed into battle. I was alone, in pain, and I was frightened.

By the second night the sharp edges of the manacles had cut deep into my back rubbing my skin raw. Every time I moved the metal dug deeper, causing pain to shoot down my spine. My feet became numb and I couldn't move my toes. There should have been a sign above the door reading, "Abandon hope, all ye who enter here," for I had certainly entered the gates of hell.

A guard periodically peered through the little window in the door to make sure I hadn't fallen asleep. If I nodded off, he came in and prodded me with his rifle, yelling in his fractured English, "No sleep. No sleep."

While at the Air Force Academy I was subjected to a quasi-torture session called the "Little Invisible Green Chair" exercise where I was instructed to press my back against the wall, slide down until my legs were bent at a 90 degree angle while holding my hands straight out in front of me. I was made to stay in that position for hours. It didn't take long before my legs and arms began to ache, feeling as though they would fall off.

That practice chair exercise was supposed to prime me psychologically for anything the enemy could throw at me, but it didn't prepared me for the relentless and agonizing pain. Here the torture went on for five days with no food, only an occasional sip of water and no way to relieve myself. All I could do was lie in my own filth with hunger biting at my stomach.

It would have been easy to give up, but I wasn't about to let the bastards win. There was no giving up. No matter how long they kept me there, I would never dishonor my country by giving them the answers they wanted.

By the sixth day I was hanging on the fringe of reality and began hallucinating. I saw myself in a big tube with squares, circles, cubes, trapezoids, and whirligigs of all different colors and size coming at me. When one image faded, another even more terrifying took its place. I tried knocking them out of my way, but they kept coming and coming. Then a raging ball of fire began following me and trailed me everywhere. The more I tried to outrun it, the closer it followed.

That's when I knew I couldn't take anymore. The next time the guard came in again and asked if I was ready to talk, I opened my parched lips and whispered, "Yes." I was ready to go beyond name, rank, and serial number.

He released me from the manacles and foot locks. When I sat up, the room began to spin. My back felt as though a knife had slashed it open and my wrists were ringed with raw, bleeding sores. I attempted to massage my fingers but they had become numb from lack of circulation. As the flow of blood slowly returned, pain throbbed in each finger making them feel like pin cushions.

When a guard handed me a jug of water, I dropped it. That was a bad thing to do. He slapped me across my face for destroying government property and for showing a lack of gratitude. I looked at the pool of water on the floor and almost cried.

I expected him to begin bombarding me with questions immediately, but instead he placed a pen and piece of paper next to me on the concrete slab. "Now, write answers," he said, and then left.

I stared at the paper as bile rose in my throat. Giving them information went against everything I believed in. Even though I felt disorientated, I was still rational enough to hold to the Code of Conduct and not give them anything that would endanger other pilots. I would go home with honor or die trying, so I decided to fabricate lies.

I tried picking up the pen, but my swollen fingers wouldn't bend to hold it. All I could do was sit looking at the paper. When the guard returned and asked, "Why you no write answers?" I shook my head and told him I couldn't hold the pen. He must have thought I was refusing to cooperate, because he beat the crap out of me again and left me lying on the floor too weak to get up. After an hour the man who had questioned me earlier, and who I assumed was an officer, came in and asked in English why I didn't write.

"I can't pick up the pen," I told him.

That answer must have satisfied him because I was pulled to my feet and taken to a different cell. In a somewhat more relaxed mood—his, not mine—the officer began questioning me again and wrote all my answers in his little blue book.

He didn't seem interested in tactical information such as airspeed, altitude, or the types of bombs we used; he was more interested in me as an American, and asked mostly biographical questions about my family, our income and background.

"What is your name?"

"Edward J. Mechenbier."

"What is your ethnic background?"

"German."

He looked at me and said, "Mec. Mec. No, that Irish."

With that I became an Irishman.

"What is father's name?"

"Edward."

"What is mother's name?"

"Helen."

In the middle of the questioning he switched gears and vainly asked, "Do you think I speak good English?" I wanted to say, "No, you sound like a stupid Vietnamese goon in a first year foreign language class." But, it seemed prudent not to say anything.

"How many brothers and sisters you have?"

"I have four sisters and three brothers."

He considered that for a moment and then said, "No. In your country nobody has that many sisters and brothers. Americans have birth control and Planned Parenthood. You have two children per family. Now, say again how many brothers and sisters you have?"

I ended up with one brother and one sister.

"What are names?" he asked.

I wondered what my family back in Dayton, Ohio, would think if they knew they were being talked about thousands of miles away in a prison cell in North Vietnam. Did they know that I had been shot down? Had anyone told them I was alive and taken prisoner?

My brother Jim, a year and a half younger than me, was my best friend, and Tom, four years younger, was a free spirit and in junior high school when I entered the Air Force Academy. My four sisters—Mary Lou, Jan, Gail, and Lora Jean—still played with dolls, and Chuck, the baby, was in diapers when I left home.

With all the digging for information about my family, I was thankful he didn't ask about my wife. Maybe because I was young and didn't wear a wedding ring, he figured I wasn't married. Either way I was relieved. I didn't want them to know anything about Jerri.

I had met Jerri during my senior year of high school while working as a dishwasher in the kitchen of Saint Elizabeth Hospital. I kept seeing this good-looking girl hanging around the locker room next to the kitchen. Some of the girls she worked with called her Jerri and some called her Claudia. After work one afternoon, I was helping myself to a pastry leftover in the pantry when she caught me. With nothing better to say, I asked her name. She told me that her given name was Claudia, but preferred being called Jerri. Mystery solved.

I offered her a ride home and several months later she asked me to her high school prom. Feeling as though I should pay her back, a few weeks later I asked her out—it was the gentlemanly thing to do—and then she asked me out again. I soon discovered she was a pretty cool chick and our relationship grew and we became an item.

I was working thirty-nine hours a week and didn't have time to date. The only reason our relationship blossomed was because we worked the same shift at the hospital and saw each other every day. Jerri stuck by me throughout my four years at the Air Force Academy. We became engaged in May during the Ring Dance of my last year and were married soon after I graduated.

Knowing that Jerri was strong willed and self-sufficient reassured me that I didn't have to worry about her. Having filled out paperwork giving her power of attorney and access to my pay before I left for Vietnam, I knew she would be okay financially and could make it without me.

"Answer question!" the officer yelled and banged his fist on the table, drawing my attention back to the grimy prison.

I saw no real value to his inquiry, so I made up names as I went along. He wrote down everything as if every word were gospel. Apparently satisfied with my answers, he looked down at me as though he had won a great battle. He had what he wanted, and I didn't give him anything useful; it was a win-win situation.

After the interview, a guard came in and beat the ever lovin' crap out of me again. What had I done this time?

"You have not been truthful. We will come back and ask you again."

Now, I was fearful that I wouldn't remember the answers I had given.

CHAPTER 3

In for the Long Haul

I SPENT THE NEXT FIVE DAYS SITTING AND WATCHING mold grow on the walls and wondering what might have happened to Kevin. But that all changed when I heard someone yelling, "Man in room number seven cough twice." This went on for three days. I had no idea what it meant until one afternoon while returning from emptying my latrine bucket, I looked up and saw the number seven above the door of my cell. That's when it dawned on me. Someone was trying to communicate with me.

The next time I heard, "Man in room number seven cough twice," I understood, and coughed.

A couple days later I heard the voice say, "Look for the wire in the rat hole."

I searched my cell for a rat hole and found a piece of paper attached to a wire poking through a small opening along the floor in the corner. I walked over to my latrine bucket and sat down, pretending to do my business. I knew I'd be in trouble should a guard open the little window (the judas door) in the door and see me doing something other than sitting on my bed. I watched the wire wiggle as if beaconing me. I unhooked the piece of paper and read, "Memorize this and then eat it."

Written in tiny letters on a one-inch square was a five-by-five matrix made of a block of five letters across and five letters down. A, B, C, D, E made up the top row with F, G, H, I, J on the second row; the rest of the alphabet followed the same pattern. Along the top and down the left side of the letters were numbers from one to five. The letter "C" replaced the letter "K."[1]

1. It is believed that Carlyle S. "Smitty" Harris was responsible for introducing the Tap Code in 1965, which he learned while a Boy Scout.

The Tap Code

	1	2	3	4	5
1	A	B	C	D	E
2	F	G	H	I	J
3	L	M	N	O	P
4	Q	R	S	T	U
5	V	W	X	Y	Z

A code! Someone had devised a code to communicate. I was elated. I wasn't alone; there was another man being held here besides Kevin and me. I did as instructed and memorized the strange graph and ate the paper.

It took practice and patience, but I soon learned the code and began communicating with its sender by tugging on the wire. To say "hi" I had to tug twice, pause, and then tug three times; tug twice, pause, and then tug four times. Sending complete messages with each word spelled out was time consuming, so we developed an abbreviated system. "The" became T; "that" changed to TT; "what" turned into WT; and "when" was WN. We used "G" for "ing" making the sentence, "When do you think we will be going home?" wn u tk we gog hom? It was a slow process, but it worked.

I found out that the man who had sent the code was Bob Shumaker who had been languishing in this prison for two years. TWO YEARS? I'm thinking, there was no way I could survive living like this for two years.

We "talked" by code every opportunity we got. Bob was hungry for news and wanted to know what was happening with the war.

I told him about meeting General William W. Momyer, commander of the 7th Air Force, when he came to Da Nang to address the pilots two days before I was shot down. He was convinced the war would last only a few more months.

"Don't sweat it," I said, "we're going home soon."

I have to hand it to Bob, he made a gallant effort to communicate, risking his life to do so. I didn't know until sometime later that Bob's cell was two doors down from mine, separated by an empty cell. He had to really work at threading that thin piece of wire across an empty storage room to hit that tiny rat hole twelve feet away. Perhaps that's why he was such a good fighter pilot; he knew how to hit his target.

The Golden Nugget

Not long after I got the code, a guard came into my cell and motioned for me to stand; I was being moved to another part of the prison. I was placed in a putrid cell so small that if I held my arms out to the side I could touch both walls. The six-foot-long cell was like a medieval dungeon straight out of the Dark Ages—dark, damp, and decaying. All it lacked were chains attached to the walls.

Here, they gave me two sets of underwear and a pair of long, two-toned red and grey, vertical-striped prison uniforms similar to pajamas, and a pair of rubber sandals made from tires. I was allowed out of my cell only once a day, long enough to empty my bucket and get my ration of thin pumpkin or cabbage soup with bits of turnip tops or seaweed, a piece of stale bread, with or without green mold, and sometimes a small lump of tasteless rice. The only protein I got was from unidentifiable bug carcasses floating in the soup thrown in for flavor. The Vietnamese apparently had never heard of a dietitian.

If the guard was in a hurry, he gave me just enough time to get my food, set on a makeshift table outside my cell, return to my room, and gulp down the gruel before he returned to collect my bowl.

When, and if, they allowed me to bathe, I got four or five minutes to dip cold water from a cistern in the courtyard and splash it over my body. While bathing, I sometimes noticed the guards turned away, diverting their eyes. I didn't know if they were uncomfortable seeing me naked, giving me a moment of privacy, or if this behavior was cultural. Either way I found their actions interesting.

With no contact with other prisoners, the days of isolation sitting on my concrete bed staring at the walls began wearing on me. The only sounds I heard were guards walking around in their squeaky sandals outside my cell and the occasional rattle of keys as a door down the hall was opened and closed.

About ten days later, a guard came in and said something in Vietnamese. Not understanding him, I just sat looking at him. He began shouting as if that would make his message clearer. It didn't. He grabbed my head and pulled it down, all the while screaming in Vietnamese.

Finally, a camp official, who spoke somewhat better English, came in and explained that I must bow to all Vietnamese. I had to stand and bow every time a guard came to the door, even when he opened the judas door, I had to bow. Bowing to everyone from the commander-in-chief of the Vietnamese armed forces down to the women who worked in the kitchen was demanded of me. Not doing so constituted a "bad attitude."

His visit was followed several minutes later by a guard who handed me a piece of paper. On it were the camp rules and regulations written in English. Seems rules were a big deal with the camp commander and he wanted to make sure I knew and understood them. Each rule was prefaced with the word "Criminal."

- Criminals must answer all questions asked by Vietnamese officers and guards
- Criminals must obey rules
- Criminals must give truthful answers
- Criminals must have good attitude
- Criminals must always bow to Vietnamese
- Criminals must keep room clean and neat
- Criminals must not make marks on walls
- Criminals must not make noise

- Criminals must not talk to other criminals
- Criminals will wake up and go to bed at the sound of the gong
- Criminals must get under bed when imperialist aggressors bomb us
- Criminals must say "Bao cao" before speaking
- Criminals can only go in area guards allow
- Criminals must not take anything into room
- Criminals must not communicate or look at other criminals when outside

Seemed simple enough.

I had been held captive for two weeks when I heard the rattle of the keys. The door opened and two guards came in, and, as before, signaled for me to gather my gear. I was being moved again.

Younger Years

Moving was something I had become accustomed to as a kid. Although I was born in Morgantown, West Virginia, I lived there for only two weeks before my father found work in Erie, Pennsylvania. Moves to Toledo, Pittsburgh, and Philadelphia, and then Lake Charles, Louisiana, followed in succession. I was too young to remember much about those places, but our move to Albuquerque, New Mexico, made an impression worth mentioning.

When we moved to the house in Albuquerque the yard was nothing but dirt. My father, wanting a nice green lawn, made my brother Jim and me plant grass seed, and wanting healthy grass, he fertilized it with horse manure he got from a friend who owned a ranch nearby. Failing to realize that a horse will eat just about anything green in its path, he spread the manure over the newly sewn seeds. Within two weeks we had a yard full of the best-looking weeds on the block.

Albuquerque was a place of never-ending excitement and adventure. Jim and I played cops and robbers with cap guns pretending to save the world from bad guys. We didn't have cap guns that used rolls of caps like the ones on television that produced a million shots. We were purists and had real six-shooter type revolver cap guns with round caps.

My first bicycle had the potential of literally causing a pain in my butt. Rather than buying a bike to fit my eight-year-old legs and then buying a new bike as I grew, my father bought a twenty-six inch silver Sunracer with pedals far beyond the reach of my feet. He fastened wooden blocks to the pedals and took off the seat, leaving the bare post sticking up. I rode standing up until my legs grew long enough for my feet to reach the pedals when sitting down.

My father liked to hunt jackrabbits. I was too small to hold a gun, so he used me to rouse the rabbits from their holes. He taught me to whistle by placing my two fingers between my lips and blowing. Once I got the hang of it, Dad took me hunting and I'd sit in the back of the pickup truck while he waited in the field. When he located a colony of rabbits he'd give a signal and I'd whistle. Hearing the sound, the rabbits popped up and dad would shoot. We made a good team. Even to this day, I can bring a roomful of people to attention with one loud whistle.

The desert gave a kid unlimited wide-open spaces in which to play. It also gave me allergies, which after catching a cold, developed into double pneumonia. By the time the doctors figured out what was wrong, I was so sick that the situation had become life threatening. My parents moved me into their bedroom and placed me in an oxygen tent for three weeks. I finally recovered, and for years mom gave me two malted milks a day to help me regain my strength.

When the doctor recommended that my parents get me away from the desert pollen, they took his advice and we moved to Dayton, Ohio. The pneumonia left me scrawny and it wasn't until my teen years that I reached the average weight and height of most other boys my age.

We lived in a three-bedroom house. Mom and Dad had one bedroom while my four sisters had the larger room. My two brothers and I shared a bedroom that was more like a back porch on the second floor. It was nice because there were a lot of windows that could be opened during summer, and for a house without air-conditioning that was a big deal. We shared a chest of drawers and a Fibber McGee and Molly hall closet that overflowed with everything from clothes to baseball gloves to schoolbooks.

Our kitchen table sat in a small nook beneath a window that had two exhaust fans on each side. Being the oldest afforded me the privilege of having first dibs on the prime spot at the table closest to the fans.

Like most families of that time, after years of sitting around the radio listening to *The Abbott and Costello Show*, *The Jack Benny Program*, and adventure shows such as *The Cisco Kid*, and *Jack Armstrong, the All-American Boy*, our family finally got a television. Then we kids sat watching *Sky King* and *Howdy Doody*. On Sunday evenings the family gathered to watch Bishop Fulton J. Sheen, which was always followed by praying the rosary.

Because we grew up in a time when everyone knew what their kids were doing, and because my father was a strict disciplinarian, I didn't get away with much. But Jim and I did have our dark side. We would go out late in the evening, lie in wait for the slow-moving trolley bus to pass, and then hopped on the back and pulled the trolley wires. If we did it just right we could snag them, causing the driver to stop the bus, get out, and untangle the wires. Sometimes it took him several tries before he got them reattached. Except for soaping windows on Halloween, that was the most mischievous prank we ever pulled. Compared to kids today, I guess you could say I had a rather boring childhood. That was good because there was a penalty for misbehaving, and corporal punishment was usually involved. My father was not one to spare the rod, belt, or hand when we needed correcting, but made sure we understood that his punishment was given with love.

I earned notoriety at a young age when I won a fishing contest by catching the largest catfish in a commercial fishing pond in Fairborn, and was interviewed by the local radio station. The reporter asked what technique I had used to catch such a big fish. I had no idea about technique; I just threw my line into the water and waited for the fish to swim by and latch onto the worm.

When I was nine I had another brush with celebrity stardom when a photograph of me was published in the *Dayton Daily News* showing the pitcher of the opposing Little League baseball team how to throw a curve ball. I was a pretty good pitcher at the time, had no idea how to throw a curve ball—that's what the photographer wanted—so I stood holding the ball while the other kid looked at it as if learning a great secret. The headline read something like "Two Best Pitchers Square Off in Local Championship Game."

I don't remember who won that year.

Around the age of fourteen, Ruth Hinkson, who lived next door to us, gave me her deceased husband's MacGregor hickory shaft, leather grip golf clubs. There were two woods, four irons, plus a putter in an old canvas bag. I learned what each wood and iron was used for, how to swing the clubs, and soon realized I loved the game. My brother Jim and I would ride the city bus to Hills and Dales in South Dayton to play golf as often as we could.

Like my father who held down jobs as a welder, steamfitter, and plumber, and sometimes sold real estate or tended bar at one of the local pubs to keep the family's finances afloat, I worked a lot of odd jobs to earn money to "supplement" my meager allowance. The first was at Roemer's grocery store where I stocked shelves and cleaned ashes from the old coal furnace in the basement three days a week. Then I worked as a bagger at Kroger's.

I also delivered the *Dayton Daily News* seven days a week. My route was Yale, Cambridge, and Oxford Avenues and part of Amberly Drive. Oxford Avenue was interesting because a whorehouse, complete with a red light on the front porch, sat on the corner. I probably took a little longer than was necessary as I pedaled past.

Wanting to earn more money than the paper route paid, and having heard about a caddy training program at the Miami Valley Golf Club, I signed up and became a caddy.

In 1957 I had the unique experience of caddying for William "Wild Bill" Ezinicki, a former National League hockey player from Manitoba, Canada, who had turned pro golfer. There I was, just a kid, carrying the bags of the New England PGA (Profesessional Golf Association) champion and walking the fairway with professionals in a real tournament. What a thrill. It didn't get any better than that. Sadly, Wild Bill lasted only two rounds before being eliminated in match play.

My father saw to it that sports became an important part of my growing up. I loved baseball and he loved being my coach. I can't remember ever playing a game when my father wasn't there in support. He, along with Ray McDonald, who had played in the Pittsburgh minor league system, coached me during the sixty to seventy games we played throughout Ohio each year. As a young teen my arm muscles weren't strong enough to allow me to play positions other than pitcher, but by the time I got to the Academy at age eighteen, I was capable of playing third base and catcher. I was a damn good bunter and became the team's pinch runner.

My dad taught me to drive our green 1951 Studebaker with a manual transmission when I turned sixteen. We'd go out every Sunday afternoon to Kroger's parking lot, set up two poles, and I'd practice parallel parking. I was a natural and mastered the technique without knocking over the poles or putting dents in the car. I aced the driver's test and promised Dad I would not become a menace to the drivers of Ohio.

When I needed a car for work, my father found a 1950 Pontiac Chief convertible with a straight-8 engine sitting in a woman's garage in Upper Dayton View. With money earned from my odd jobs, I bought it for thirty-five dollars. It was a rattling bucket of bolts, but I had confidence I could fix it up. The floorboard was rusted out so we put a piece of plywood on the floor and drove it to Tate Station where my father worked as a welder.

He welded a thick piece of steel to the floor strong enough that the car could have driven over a land mine without being damaged. I pounded out the dents, sanded the rust spots, and applied Bondo, and then painted the car sandstone white. Using an eight-pound test fishing line, I stitched the rips in the car seats and oiled the dry, cracked leather with Neatsfoot Oil until it softened. When I finally finished, I had the coolest car on the road.

Two weeks later I got my first speeding ticket going 35 miles an hour in a 25-mile-an-hour zone in downtown Dayton. So much for promises not to become a threat to others. I was summoned to juvenile court to stand before a judge who restricted my driving privileges for six weeks. I expected to get a good talking to from my dad about the responsibilities of car ownership and the importance of obeying the laws of the road, but he must have thought the whole episode was silly because I never heard a word. I drove that car for several years and later gave it to my brother Jim when I left for the Air Force Academy.

Where my father made time for us during baseball season, mom took time away from her household chores to attend our after school activities. Mom was a beautiful woman and I was always proud of her. She looked young and had the enthusiasm of a cheerleader. Whenever we went to watch my brother Jim, the All Star, play football, people thought she was my date.

I can still picture my mother in her favorite place in the kitchen cooking stick-to-the-ribs meals and making my favorite cherry pie. As the glue that held the family together, Mom had a knack for stretching a dollar's worth of hamburger to feed the entire family. Even with her busy schedule, she found time to starch and iron my pillowcase because I had once mentioned that I liked the way my pillowcase felt.

My mother was diligent in making sure I saved my money. Everything I earned went into a savings account. I recall walking into People's Bank with my passbook and my mom putting the money in the bank for me. It was my money to spend however I wanted, but my ultimate goal was to save enough for college.

I joined the Cub Scouts in fifth grade and became a Boy Scout when I turned twelve. I liked the structure of the organization, wearing the uniform and saluting; a precursor of events to come. Camping and being outdoors were my favorite activities. Every summer we went to the Treaty Jamboree in Greenville, Ohio, where I was an Order of the Arrow instructor, teaching younger scouts how to start a fire with friction, pitch a tent, and survive in the wild. I made both Star and Life Scout, but left for the Academy before earning Eagle Scout.

In my sophomore year at Chaminade High School, I played baritone horn in both marching and concert bands until playing such an instrument became uncool.

I guess you could say my career in uniform began as a Boy Scout and as a member of the high school marching band. Marching band taught me how to stay in line while moving my feet and blowing a horn. Scouting taught me leadership and how to survive the rigors of doing without. Who knew?

As one guard blindfolded me, another, muttering in Vietnamese, tied my hands behind my back. But instead of being led to a different cell within the prison like before, I was put into a small three-wheeled scooter and driven beyond the prison walls.

CHAPTER 4

The Power Plant

THE YEN PHU POWER PLANT, located in the industrial center of the northern part of Hanoi, was comprised of a series of buildings spread over five city blocks and included a machine shop, warehouse, underground factory, a school, market place, and a medical aid station. The complex was filthy and covered with a layer of soot from the factory, earning it the name Dirty Bird.

I was taken to what could best be described as a room in a single-story fleabag motel-style building that faced onto a busy street. As dreary as the place was, it was still better than the dark and depressing cell where I had previously been held. The one small, boarded window through which I could see people walking past, allowed sunshine to seep between the cracks during the day and a faint breeze to filter in at night. Through the cracks I could watch civilians walking past. Several times, I saw A-4 Skyhawks overhead, dodging barrage balloons suspended two or three thousand feet in the air.

On the backside of my cell was an interior corridor where guards set out my meal of hot, sweetened brown rice, the consistency of pudding, which provided my total sum of calories every day. Just beyond the corridor, a brick wall partially hid the remains of a bombed-out power plant.

The days were uneventful and I was spared the endless hours of interrogation. No beatings. No shouting. Nothing. More importantly I no longer feared for my safety. Compared to the "attitude adjustment sessions" I received while at Thunderbird 7, life here was bearable. Of course, the camp rules still applied.

Even weeks after being released from the foot locks my ankles were still raw with sores. I was in constant pain from where the manacles had gnawed into my back. All I could do was lie on my stomach or side. I had little use of my hands and couldn't hold a spoon and had to palm my bowl to eat the gruel they provided.

Every morning after my first meal, a guard directed me to put on the two-tone long sleeve shirt and pants I called my mess dress uniform. He blindfolded me, and then took me for a ride that ended at an underground factory not far from the power plant. I found it interesting that sometimes the trip took longer than other times. I'm guessing he didn't want me to know exactly how close the place was to where I was being held, but I suspected the building was along the river because I could hear the sound of boat motors and splashing water.

I was taken down a flight of metal stairs and chained to a railing on the landing. There, I was made to stand for five to six hours like a rack of beef on display in a butcher shop. Workers trekked up and down the stairs. Some walked past, taking little notice of me, some were kind and tried slipping me cigarettes and candy. Others looked at me with contempt as though they would rather kick me than offer any act of human kindness. A guard, with a forlorn expression sat watch over me to make certain I didn't try to sit down or talk to anyone. He looked as unhappy as I felt.

I was given a sip of water occasionally, but going to the bathroom wasn't an option. Although my back was killing me, I stood there. What else could I do tethered as I was?

The Vietnamese never explained why I was taken to that underground stairway. Perhaps my presence was to reassure the general population that the North Vietnamese held captured American prisoners. Maybe I was there to prevent the Americans from bombing the place. It wouldn't have been unlike the Vietnamese to send word to our government saying they were holding prisoners near strategic targets, and if the United States bombed the industrial center they would be killing Americans. I don't know if that were true, but it made sense to me.

Standing there those many hours chained to the railing gave me plenty of time to think about what was going on in Washington and what our leaders were doing, if anything, to end the fighting. Obviously General Momyer's prediction of only a few more months of war was wrong, leaving me to wonder how much anyone really knew about how long I would be held prisoner.

It would have been easy to fall into the trap of self-pity and become melancholy, wishing none of this had happened and imagining myself at home, but I never allowed myself to think about anything except the present. I had to stay focused on the here and now. I had been taught at the Academy that fantasizing about what could have been or wishing for something that wasn't at hand could only sink a person into depression. So I stayed in my skin and didn't think about what ifs. Just trying to cope with the horrific everyday events took all my energy, and I needed every bit of strength I could summon to stay alive. I was going to hold on and not let the goons get the best of me. Just as it took skill, fast reflexes, and airmanship to be a good fighter pilot, I discovered that being a prisoner of war took tenacity, cunning, and dogged determination to survive.

The $5 Bet

I suppose I could have blamed my father for my current predicament. If I hadn't lost a $5 bet to him, I might not have ended up where I was.

Raising eight kids took every cent my father made and with money always tight, he told me that if I wanted to go to college, I'd have to earn a scholarship. After seeing an article in the local newspaper announcing that Congressman Paul F. Schenck of Dayton was conducting interviews for appointments to the Air Force Academy, he suggested I apply. I had no interest in the Air Force or airplanes, but my dad had faith in me and bet me five dollars I could earn an appointment. So, I took him up on the bet certain the five dollars would be in my pocket.

I went through the process, first writing a letter to Congressman Schenck, requesting an appointment, and then underwent an initial interview with him. Finally, I was invited to Wright-Patterson Air Force Base (WPAFB) to take a physical to make sure I was fit to serve.

Doubting that I would get an appointment, and thinking that I wanted to be a lawyer, I applied for a needs scholarship through the high school counselor's office and made plans to go to Saint Joseph College in Kansas City, Missouri, to study law, but that all changed on May 13, 1960, when I got a letter of acceptance to the Air Force Academy.

The Academy sent a list of items I could and couldn't take with me; one was money. Despite having saved six hundred dollars working those many odd jobs as a kid, I couldn't take it, consequently, the day I left for the Academy I had to borrow five dollars from my mother at the airport to pay off the bet to my dad.

My very first airplane ride was a five-hour flight to Colorado Springs in a TWA Super-G Constellation. I arrived at Peterson Field on June 27 and was met at the airport by a bus that carried me to the Air Force Academy. Even before I was off the bus, a cadet from the basic training detail began barking orders to drop and give him twenty pushups. I got a close-up view of Colorado with my nose six inches from the ground.

Hello, Colorado.

Along with other newly arrived cadets, I was barely given time to unpack or become acclimated to the altitude before being shepherded to the barbershop for a little trim. "Stay in line, eyes straight ahead, no talking," ordered the senior cadet as I stood waiting my turn in the chair. I went into the barbershop a normal teenager with hair hanging down over my forehead and came out a shaved melon head.

The second day I raised my hand and took the enlistment oath that said, "I will support and defend the Constitution of the United States against all enemies, foreign and domestic" and became a cadet, a rank between that of a master sergeant and a second lieutenant in the Air Force with half the base pay of a second lieutenant. I was now subject to the Uniform Code of Military Justice, which meant that if I screwed up, I could be charged for my offense, brought before a board, tried, and, if found guilty, punished.

So much for college fun.

On the third day during a "get acquainted" session, all the new cadets of class 1964 were marched to the auditorium. The class leader said, "All of you who were high school class presidents, stand up. All who were captains of your football team, stand up. All who were voted Most Likely to Succeed, stand up."

This went on for several minutes until he had exhausted all categories for defining success. I looked around and saw that I was one of only a few men still seated among a group of highly successful and motivated young men who most likely got to the Academy because of their scholastic achievements, leaving me to wonder how in the heck I had won an appointment. All those "up standing" achievers had done something special during high school and I was there because I lost a five dollar bet. Boy, was I in way over my head.

The first year was intense, both physically and emotionally. As underclassmen, we marched everywhere and were required to square the corners and stay on the north/south, east/west marble strips of tile on the Terrazzo. We were often deprived of sleep, endured repeated hazing, and withstood personal attacks because of our accent, hair color, stature, and parentage. The purpose of the training was to bring everyone down to a common denominator so they could build us back up as a cohesive unit. There was no room for individuality.

Our day started at 0600 hours. After a quick breakfast it was off to class for indoctrination into the military. I was required to read and memorize everything within the one-hundred-page manual of *Contrails* and be prepared to recite passages whenever tested by an upperclassman. I was expected to be able to identify military aircraft, memorize quotes from famous generals, the major air commands, and the names of just about everyone associated with the Air Force from the chief of staff in Washington down to the cadets in charge of my dormitory floor. That was a lot of information to cram into my brain.

At 1800 hours we lined up for dinner formation, followed by night classes. When taps sounded at 2200 hours, we got out our flashlights and pulled the blankets over our heads to study.

Personal grooming inspections meant a close shave, fingernails trimmed and clean, and a straight gig line on our uniform. Even one hair out of place was cause for a demerit. We spit-shined our shoes until we could see our reflection in the toes.

Our rooms and common areas were inspected daily. My bed had to be made with squared corners and the blanket pulled tight enough that the inspector could bounce a quarter off the center. We had to fold our underwear and socks to exact specifications, and hang our razor-sharp creased uniforms in the closet facing one direction. I looked upon that closet with pride; I never had one as a kid.

If I failed to correctly quote information from *Contrails*, I received extra demerits. If I screwed up, I'd have to do it over and over until I got it right and sometimes that meant missing a meal.

I became the personal project of Richard D. Smith, an upperclassman two years ahead of me. He seemed to go out of his way to point out my failings and made sure I was given extra details and material to study that wasn't always in *Contrails.*

He stayed on my case and Dumbsmack and Doolie were only a few of the names he called me.

"Do you want to be here, Doolie?" Smith yelled.

"Yes, sir."

"Do you miss your mommy, Dumbsmack?"

"No, sir."

He even had me quote from *Contrails* while looking at a picture of a shapely young woman wearing a teeny weenie yellow polka dot bikini. "What are you looking at, Doolie?"

"Ah, Ah. I don't know, sir," I said, looking at the picture only inches from my nose.

"Are you laughing? Are you smirking, Dumbsmack?

I stood ramrod straight and shouted, "No, sir."

It was hard not to burst out laughing. Why he picked me out of all the cadets to focus his attention, I don't know, but he seemed determined to get me booted out of the Academy. His constant hounding only made me all the more determined to do well and show him that I could make it through. Perhaps that was his hidden agenda all along.

I carried my M-1 rifle, serial number 5770855, all through basic training; it became an intimate friend. I had to keep it spotless, know all its parts, and be able to take it apart and reassemble it blindfolded. I was told that rifle could someday save my life.

Physical training consisted of mandatory classes in Judo, handball, squash, boxing, swimming, and wrestling. It helped that I was in good physical condition from playing baseball and caddying and had developed upper body strength. That made life a lot easier when it came to doing pushups.

That same routine continued for eight weeks with one big exception—I got to ride in a Lockheed T-33 Shooting Star trainer airplane for a one-hour orientation flight. I had no idea what flying in a jet fighter was all about and felt apprehensive. We were given a brief familiarization class on the aircraft, shown how to strap on a parachute and use the ejection seat should it become necessary. The instructor made it very clear that I was not to touch anything because doing so could cause the airplane to crash. He told me that if anything went wrong, he would bang on the canopy three times. That meant we were going to eject. After all the preparation, and with the fear of God in me, I wasn't exactly giddy with excitement about getting into the airplane and leaving solid ground, but I climbed into the back seat, strapped in, and was taken for the ride of my life.

Once past the hostile, threatening, and scary environment, I relaxed and began to enjoy the ride. The pilot flew the airplane upside down and, despite his earlier prediction of doom if I touched anything, let me work the controls for a few minutes. There I was, eighteen years old flying a jet fighter. From that moment I was hooked and knew I wanted to be a pilot.

In the fall of that first year, while playing Lacrosse, I took a serious hit to my manhood. While upperclassmen got the plush scoring positions, newbies were relegated to the back of the field to cover defense where we got knocked around a lot.

I was standing between the opposing team's offense line and the crease, the circle around the goal, guarding against a high shot when the shooter came in with a low swing. Before I could lower my stick, I got drilled; I mean that ball slammed into my testicles like a rocket. Although I had on kneepads and a helmet, I was not wearing a protective cup. The initial pain was excruciating. I cried out an octave higher than was acceptable for a big, strong Air Force Academy cadet, and suffered the humiliating ribbing of my teammates. It was late in the second half, and after regaining a somewhat upright stance, I stayed in and finished the last few minutes of the game.

Needing showers, the team ran the mile from the athletic field back to the field house. I made it half way before passing out.

I regained consciousness stretched out on a bed with doctors and nurses staring down at me. The official diagnosis was severe contusion of the testicles. I was transported to the cadet dispensary where I spent the next three weeks packed in ice to relieve the swelling in my groin. The doctor ascertained that the impact of the ball hitting me with such force triggered intense internal convulsions that caused my intestines to become trapped in my stomach muscles. It not only took time for the swelling to subside, but it also took time for all my innards to return to their proper place.

It was just my luck to have Dave Bushman, the 1963 class clown, who was laid up with a broken arm, for a hospital roommate. Being a second-year man, he was afforded privileges, and when two young ladies came to visit him he used me as his second banana. "Ask Mechenbier what his problem is," he said to the young ladies.

I wasn't thrilled lying flat on my back with a mound of ice covering my lower half while he made jokes about my condition. It hurt to move much less laugh. He and I actually became good friends after Hell Week and Recognition when I was free to socialize with upperclassmen. That's when I was accepted into the cadet wing, and earned a cadet prop and wing and was no longer considered a one-cell life form.

In the fall of my second year I took a hard hit to my left knee while playing intramural football and tore the meniscus, the tissue between the tibia and the femur. To fix the problem an orthopedic surgeon, who had the reputation of "practicing" on his patients, made an eight-inch incision on the side of my leg, but instead of going in and taking out the damaged pieces of the meniscus, he took a chisel, chipped the ligament off the bone, and then took out the patella. I was awake during all this, having been given a spinal block, and I heard him say, "Oh, all you have is a torn meniscus. We didn't need to do all that work."

To reattach the ligament, he drilled a hole in the bone and inserted a one-inch Vitallium boat nail to hold it in place.

It was bad enough that I was sidelined from classes for a month because of the injury. I was also failing physics and astronautics with a solid F. Most educational institutions would say sorry, and let a student sink to the bottom, but that wasn't the Academy's way. Two wonderful instructors came to the hospital, tutored me, and truly saved my career. I ended up earning As and Bs in those classes.

One of the most important lessons I took away from my first year at the Academy was the feeling that everything the instructors did was to ensure my success. Beneath the impersonal military structure, there was a sincere dedication on the part of all the instructors to see that every cadet succeeded in graduating. It was awfully damn tough, but because of their commitment I never, ever, doubted I was going to make it.

As part of my training during the summer between my second and third year, I was given the opportunity to travel. Most everyone wanted to go to Europe, and because I felt I would never have the chance to see the Orient, I chose the Far East and went to Japan, Okinawa, Thailand, Hong Kong, Philippines, Guam, and Hawaii. What a great experience that was. I also got to ride in a TF-102 Delta Dagger, an air-to-air interceptor, that gave me the feeling of what it was like to be a fighter pilot. I was literally vaulting through the air rather than merely passing through it. That was the defining moment when the barb on the hook was set. I knew I didn't want to be just a pilot, I wanted to be a fighter pilot.

In January 1961 the Academy was informed that the entire cadet wing was to participate in President John F. Kennedy's inauguration parade in Washington. We made an extra effort to look our best with spit-shined shoes and impeccably pressed uniforms before being loaded onto an old, noisy, unheated, three-bladed prop C-130 Hercules where we sat in four rows of troop seats facing each other. The weather forecast predicted snow and rain with bitterly cold temperatures all along the East Coast, making a landing at Andrews Air Force Base questionable. The first few aircraft that reached Washington were allowed to land, but by the time we arrived conditions had fallen below ceiling and visibility minimums, and we were turned away.

We were freezing as the airplane was redirected to Fort Campbell in Kentucky sometime around 0200 hours. Cold and hungry, the Army fed us hash and billeted us in an old, drafty World War II barracks. Just as we began to thaw out we were loaded back onto that same cold and noisy airplane for another try at Washington. The landing conditions hadn't improved and we were forced to return to Colorado in the same airplane. I can't remember ever being so cold.

Of the fifty airplanes used to fly the twenty-four hundred cadets to Washington, only twelve made it. The rest of us sat in the day room of Vandenberg Hall watching our fellow cadets on television parade through the streets of Washington looking like frozen walking sticks in their blue horse-blanket topcoats.

Sadly, several years later we sat in that same day room watching in shock and disbelief listening to the cadence of muffled drums and the clacking hooves of the riderless horse, Black Jack, echoing across the nation as President Kennedy's funeral cortège made its way down the street to the Capitol.

In June 1963 at the end of my third year, Jerri and I became engaged. She came to Colorado Springs that fall and lived in a convent at Penrose Hospital with her aunt, who was a Charity nun and surgical nurse. We could see each other only on weekends, and sometimes not even then because of some dumb thing over which I had no control, but for which I had total responsibility. In other words, I screwed up and was on occasion confined to my room at the Academy while my fiancée sat thirteen miles away in a convent.

During the summer of 1963 between my third and fourth year, I was assigned to an operational squadron with the 916th Air Refueling Wing at Travis Air Force Base in California as part of the program called Operation Third Lieutenant. During the five-week training program I got to ride in a KC-135 Stratotanker. That experience convinced me that I definitely did not want to become a trash hauler or gas passer.

After four years I had racked up 209 credit hours and graduated with a bachelor's degree in Engineering Science and was commissioned on June 3, 1964, earning my butter bars as a second lieutenant.

Mixed emotions tugged at me as I prepared to leave. Looking around, I knew that everything I had gone through over the past four years, all the hard work and diligent attention expected of me as a cadet, was coming to an end. The Academy had given me an education and molded me into a man. Although I was excited and looking forward to a career in the military, I was also sad leaving a place that had become a comfortable womb.

CHAPTER 5

Flight School

FOLLOWING GRADUATION FROM THE ACADEMY, I was given the choice of attending the flight training class that started in two months or one starting ten weeks later. Eager to begin learning to fly, I took the first class available—Class 66A.

Meanwhile, Jerri and I got married and raced through the Smokey Mountains and bolted past New Orleans in a drive-by honeymoon. We found a little house near Vance Air Force Base in Oklahoma and settled down as a young married couple before I reported for basic flight training.

Entering flight school, I was like a kid at Christmas with the tree surrounded by new toys. I was excited. I was hyped. I was going to fly a real airplane. But those first few weeks of training, from my perspective, were anything but exciting; they were downright boring.

We spent weeks learning the basic principles of aerodynamics and how the forces of lift, drag, weight, and thrust affected an airplane in flight as the instructor talked about the primary movement of roll, pitch, and yaw, and then droned on and on about the lateral, vertical, and longitudinal axis. I already had two years of astrophysics and aeronautics and here he was trying to teach me the basic theory of flight. I knew how an airplane worked. Why was I wasting brain cells on this? GET ON WITH IT! I wanted to get in the cockpit and go fly.

After staring at books and manuals filled with charts, diagrams, and regulations for three grueling weeks, I finally got what was called a dollar ride in a duel-engine Cessna T-37 Tweet. The T-37, with its Continental-Teledyne J69-T-9 Turbojet twin engines, was used as the primary training aircraft to introduce neophytes into the world of fighter jets. The side-by-side seating allowed for better interaction between the student and instructor as opposed to the front-to-rear configuration. The Tweety Bird, as it was affectionately called, had low, straight wings, a tricycle landing gear, and gave off a high-pitched, ear-splitting shriek that some compared to a dog whistle.

Captain Ed Eubrecht, my instructor pilot, was a taskmaster and the reason I did so well. He taught me the discipline necessary to become not just a pilot, but a damn good one. He wouldn't accept that close was good enough. Each maneuver had to be precise. One or two degrees off here, a couple miles off there were unacceptable.

During that first ride I sat beside him while he flew the airplane. The purpose of the flight was to see how I would react emotionally, psychologically, and physically to the stresses of flying as he put the airplane through the various maneuvers of rolls, spins, and stalls. Would I become disorientated and get airsick? Luckily, I did just fine. That flight was an absolute blast; better than any ride at Disney World.

During egress training to see how I handled g-forces, I sat on an ejection seat that propelled me thirty feet up a rail to give me the feel of what it was like to eject from an aircraft. Three years later I discovered that training had no similarity to a real ejection.

Solo Flight

After thirteen flight-training hours, which was about the minimum, I had fulfilled all the requirements and was ready for my first solo flight.

I had already made several practice landings, when Eubrecht unexpectedly cut the training session short and told me to make the next landing a full stop. I figured I had screwed up and would be admonished for some infraction. We taxied to the end of runway approach. Eubrecht got out of the airplane, turned back, and then said, "Go shoot three landings. See you later."

Nervously, I taxied to the end of the runway, pushed the throttles forward, and "slipped the surly bonds of earth" as John Gillespie Magee, Jr., wrote in his poem "High Flight." Once in the air, I looked over at the empty seat beside me. Wow! I was finally doing something I had only dreamed of; I was flying an airplane by myself.

I made two "touch-and-go" and one "full-stop" landing, and then taxied to the squadron area and climbed out of the airplane. My fellow student pilots, keeping to the tradition of the first solo wet down, picked me up and threw me into the water trough. I climbed out soaked, but happy. I was finally a real pilot.

I was excited to be the first in my class to solo, but to bring me down a peg Eubrecht got one over on me. On my next flight, after flying for about a half hour, he told me to bring it home. I came in on initial at 200 knots in a clean configuration with gear and flaps up, pitched out in a 60 degree bank, made a 180 degree turn, pulled the throttles back to idle to lose air speed, and then continued rolling out on downwind. When the air speed got below the maximum allowable gear-lowering limit, I waited for a chance to announce "gear down," but Eubrecht started talking, telling me that I was absolutely the most fantastic student pilot he'd ever had. "You have picked this up in a hurry. You know this business backward and forward. You have good situational awareness. You'll probably graduate top in your class," he said.

He went on and on showering me with praise, not allowing me to lower the gear without interrupting him to say I was changing the configuration of the airplane. I came around on final approach with the gear up and got the red warning flare. How embarrassing! When we finally landed, he started laughing. "You needed that," he said with a big grin.

I had another superb instructor by the name of Major Archie Clarke, a former combat pilot in Korea. During my training, whenever I started overworking the controls, he let me make mistakes, and then figure out for myself how to correct them. He never screamed or jerked me around by the oxygen mask's hose as some instructors were known to do. Archie taught me how to fly by keeping it simple and fun. I owe him a lot.

Following the fifty-four-week flight training program at Vance, I graduated on August 4, 1965. I had high hopes of becoming an F-100 Super Sabre pilot, but no F-100s were available in that assignment block. Despite having ranked top in my class, I had to take what I didn't want—flying an F-4, which meant sitting in the back seat while someone else commanded the airplane. With no other choice, I accepted a three-year assignment to England with the 92nd Fighter Squadron, 81st Tactical Fighter Wing at RAF Bentwaters, eighty miles northeast of London near Woodbridge, Suffolk.

With the ever-present danger of being shot down and taken prisoner, pilots were required survival training. So before going to England, I was sent TDY (temporary duty) to Stead Air Force Base in Nevada while Jerri returned to Ohio. At Stead I spent many hours in a simulated Vietnam prison camp where I was tested mentally and physically. I was exposed to a harsh, cold, wet environment, interrogated, harassed, and verbally assaulted while instructors played the role of good cop, bad cop. Although I was never physically abused, there was psychological deprivation. I was isolated in a small, dark room with constant white noise and given nothing to eat. The training was so intense that it was easy to forget that everything I was going through was only an exercise. There was still a lingering uncertainty that maybe they were playing for keeps and I might not walk out of the room alive; the instructors were just that good.

When survival training ended, before heading off to England, I was given orders to Davis-Monthan Air Force Base in Tucson, Arizona, for Combat Crew Training School where I gained experience in the intricacies of the McDonnell-Douglas F-4C/Phantom-II. Part of that training consisted of learning how to deliver nuclear weapons with a couple thousand pounds of concrete attached under the airplane's wing to simulate the aerodynamics of the B-57 and B-61 bombs.

As a young, freshly minted, second lieutenant fighter pilot, I thought I was hot stuff, but got a dose of reality when, as a GIB (Guy in the Back) I had to work with more experienced captains and majors. Having always been single-seat fighter pilots, they weren't too keen on having someone sitting in their back seat sharing duties, responsibilities, and accolades.

Social hour at the bar was awkward. They tried to be nice, but it was obvious they'd rather be with men of their own kind. As a backseater, I was the young kid on the block trying to emulate the macho fighter pilots. They, on the other hand, had established their credentials, earned their swagger, and drew a line to make it clear I hadn't yet paid my dues.

After graduating from Combat Crew Training School in February, I was assigned as the GIB, to fly with a pilot-in-training I'll call "Major Incompetent" who, in my opinion, was unskilled and not fit to fly a complicated and expensive aircraft like the F-4C, but because of his rank as a major, he was placed in the position of aircraft commander. Every landing with him was a dance with death, even though the F-4 is unquestionably the easiest airplane to fly.

The squadron in England was still flying F-101s but was in the process of transitioning to new F-4s, so as each new crew finished training, they were detailed to the factory at Saint Louis to pick up a new F-4 that would become part of a newly formed squadron. Major Incompetent and I picked up a brand new aircraft at the factory and then flew it to Robins Air Force Base in Georgia and waited for three more airplanes to come off the production line before linking up with a tanker to high-flight the aircraft across the ocean to Europe.

Getting into an F-4 can be awkward, especially with your hands full. Carrying my personal gear along with the airplane's papers, I climbed the ladder and stepped onto the left engine to reach the back seat. I set my Dopp kit containing all my toiletries and the airplane's 781 forms on the flat top of the engine's intake cover and climbed into the cockpit. The crew chief strapped me in and after finishing the preflight procedures we took off. It wasn't until we got to Torrejon Air Base in Spain that I realized I had left the forms and my Dopp kit sitting on the engine.

Some weeks later the forms, along with my Dopp kit, arrived in somewhat less than usable condition. Apparently, they had blown off and were crushed under the tires of the next airplane taking off.

But forms and Dopp kit were the least of my concerns. It was a miracle that I made it to England at all. After a stop in Spain, we headed to RAF Bentwaters on the last leg of the flight and were at twenty-five thousand feet when I noticed Major Incompetent had forgotten to pressurize the airplane, and I started to show symptoms of hypoxia. When I told him that the cabin wasn't pressurized, he shoved the lever down. POW! The cabin became instantly pressurized.

Assignment to England

The military had strict rules about wives accompanying their husbands on overseas assignments, and they had to wait until quarters were secured before being allowed to join their husbands. That process took about a month and when Jerri finally arrived in England we found a house called Pear Tree Cottage on the Foxboro Hall Estate, a place with a recorded history dating back to the 1600s.

Our landlord, Major W. E. P. Miller, had been the commander of the Royal

Guard. We figured he must have been someone of importance when we saw a photograph of him standing close to Elizabeth II as the Archbishop of Canterbury crowned her Queen of England on her coronation day.

Major Miller had a long list of people wanting to rent the cottage but Jerri turned on her charm and we got the cottage for seventy-nine dollars a month. The British conduct business in a grand manner. Our lease was a twelve-page bound document, handwritten on parchment with the barrister's seal on the cover, and my name read as "Lieutenant Edward John Mechenbier, Esq."

The solid stone house had a bath upstairs and a fireplace in each of the four bedrooms, which turned out to be crucial because the house had no insulation or central heating. One corner of the house had been bombed during World War II and the wall replaced with brick and stucco, making it appear to be stone. Downstairs there was a nice size living room, dining room, and kitchen with real worm-eaten furniture. Ceilings and doorways reflected the short height of the people at the time the house was built, which meant Jerri and I, both being rather tall, had to duck to walk through.

The Queen visited the estate once a year to view the rose gardens and the holly bushes in our yard and we were given instructions not to be home when she made her appearance. What an honor it would have been to see the Queen of England in our front yard. That was as close as I ever came to rubbing shoulders with royalty.

Jerri and I toured as much as possible seeing the sights and visiting castles, such as Orford Castle in Suffolk. After one of our outings, we received a written summons to appear at the manor house at Foxboro Hall.

We arrived at the appointed hour and were shown into a sitting room where the major greeted us warmly. After exchanging pleasantries, the equivalent of the American "hello," he turned to me and asked, "Lieutenant, what did you do this past weekend?"

"Well, sir, we had a wonderful time. We went to Framlingham Castle."

He sat back, tented his fingers and said in his most stiff, upper-lip British accent, "Ah, yes, yes, Framlingham. Grand old place. And what did you do there?"

"We wandered down into the village, window shopped and visited a pub."

"Ah, yes, yes, a pub. Yes, you did."

Then he leaned forward and gave me a steely-eyed stare. "Surely you are aware that there are two parts of any pub."

"Yes, sir," I said, wondering where he was going with this.

"There is the public bar and there is the private bar. It has been reported to me that you were in the public bar."

"Yes, sir," I said, still curious as to what he was getting at.

"My secretary owns that establishment and reports to me. You are an officer and are not to go into the public bar. You will go to the private bar. Upon your next visit, identify yourself and be invited to the private bar."

That was the total purpose of our summons to the manor house.

Upgrade to Aircraft Commander

My first few months in England were filled with intense training, flying low-level navigation missions and dropping practice "tactical nukes" on sites in Western Europe. Flying with the same incompetent aircraft commander, it became more and more apparent that he was dangerous in the cockpit. I went to the operations officer, Lieutenant Colonel Dana Borne, and said that I refused to fly with the guy. I didn't want to be in the aircraft when he came to the end of his competency and took me with him.

Being reassigned as a backseater to a new aircraft commander was a relief, but I didn't know I was about to go from the frying pan into the fire. Only this situation had a more positive outcome.

Most of my training while stationed in Europe took place at Wheelus Air Base near Tripoli. Wheelus was where I first experienced the ever-soft gentle breeze called Zephyr. In a land of French Moroccan architecture, date palms, white sandy beaches, and the blue of the Mediterranean Sea, Wheelus was unlike any place I had ever been. Built in the early 1900s by the Italian Air Force, the base still had World War II vintage buildings. Temperatures reaching into the 100-degree range and the frequent dust storms made training unique. It wasn't uncommon to see camels plodding along at the end of the runway.

Now flying with another aircraft commander, I spent several weeks in the back seat practicing dropping bombs at the El Watia Gunnery Range. While flying a training mission, making a 45-degree dive bomb at ten thousand feet, the pilot, who I will call "Captain Freeze," lost control of the airplane and put us into a spin. The SOP states that if you are below eighteen thousand feet and the airplane goes out of control you eject, but the pilot froze, leaving the aircraft spiraling toward earth.

Realizing that Freeze wasn't responding, I yelled through the headset, "Tom, I think you've spun the f——ker. Pull the drag chute."

I repeated the command. He finally pulled the chute, but by that time we were in serious trouble. Seeing that he wasn't going to take control, I grabbed the stick, kicked the opposite rudder against the spin and with ailerons in neutral, locked the stick between my knees to hold it steady, pitched the nose forward to gain air speed, radioed the tower and declared an emergency. I brought the airplane back to the base, having missed the ground by only twenty feet and whipping up dust devils as we barely recovered in time to make a safe landing.

Captain Freeze tried to tell me that the aircraft had a control malfunction. I knew there had been no malfunction because whatever could have been wrong with the controls didn't manifest itself during the eighty-mile flight back to the base.

A series of investigations followed. Although I knew what the pilot had done, I never told anyone. When asked, I told the investigators that I had flown the airplane back and landed it, and if they wanted to know the details, they'd have to talk to the aircraft commander.

Investigators took the airplane apart trying to find anything that pointed to a mechanical problem. A week later a McDonnell tech rep came up to me and said, "I've been in this business for thirty years and I have never not been able to find a mechanical problem with an airplane that kept a pilot from being tagged as a screw up. There was nothing wrong with that airplane, so let me tell you what happened," and then he proceeded to describe the series of events giving the air speed, altitude, number of rotations, second by second, precisely as they occurred. Then he asked me if he was correct. I told him that I wasn't in the position to say.

The next day Brigadier General Dave Jones from USAFE (U.S. Armed Forces Europe) mentioned the incident and told me he appreciated how I had handled the situation, and then said, "Oh, by the way, you're now an aircraft commander."

I had a one-star general approve my upgrade.

A young pilot can spend an entire tour in the back seat of an F-4, which could last as long as three or four years or about eight hundred to one thousand flying hours. But with less than one hundred hours in the back seat I became the only first lieutenant aircraft commander flying an F-4 Phantom in Europe. I was a little self-conscious because there were guys who were senior to me who were still flying in the back seat.

Even though I was officially an aircraft commander, I still needed further training. As a rule, a pilot is sent to Davis-Monthan for the aircraft commander check out, but I stayed on at Bentwaters where I acquired the skill set needed to fly the airplane from the front seat. J.J. Davis, another classmate from the Air Force Academy, was assigned as my first GIB. He was a good pilot and, as a team, we learned the art of being a combat crew.

With offensive nuclear alert as part of my job, at 0800, an hour after taking my final check ride in the F-4, I sat down before the Combat Crew Certification Board consisting of three lieutenant colonels, one from operations, one from the nuclear weapons shop, and someone from the plans shop for the verbal test that would qualify me to sit on alerts. I was asked to explain the key elements of the profile, including timing, waypoints, conflicts, launch points, air speeds, and egress maneuvers after delivering a nuclear weapon. They tested me on what kind of targets we were looking at, whether it was a defensive site, airfield, or industrial complex, and what, if any, civilian casualties there might be as a result of my bombs.

I must have had the correct answers because they said, "Congratulations, you're certified, now get on the bus."

They really needed pilots for standby nuclear alert and needed them by noon.

Blue Bark Duty

I had been in England only a short time when I had my first experience with a pilot's death. Tom Moore, an F-4 backseater, and his aircraft commander, Len Toverie, crashed into the North Sea and were killed. Their deaths were an awakening. For the first time it fully sunk in that men could actually die doing what I loved, and I began to understand and appreciate the dangers of my chosen career. That was a sobering moment.

The squadron commander, Colonel Slade Nash, tasked me to accompany him and the chaplain, along with a doctor to give Tom's wife, Prissy, the news that her husband had been killed. Fortunately, all I had to do was introduce the men with me. I didn't have to actually say the words, "Your husband is dead." I was, however, tapped with the Blue Bark duty of escorting Tom's body and his wife and son back home to a little town outside Little Rock, Arkansas.

While waiting at JFK International Airport for our flight connection to Arkansas, a Pan Am worker came up to me and asked if by any chance I spoke German. How he could have known that German was spoken in my home and was the native tongue of my parents was beyond me? I told him yes, and he asked me to follow him. He directed me to a woman fitting the description of the typical older matronly German lady, dressed in a tweed suit, matching hat, and carrying a handbag over her arm.

The Pan Am representative told me he had been trying to make the woman understand she had only five minutes to make her connection to Detroit, and needed to go to the other gate immediately. Not understanding English, she refused to budge.

Speaking in German, the woman, told me she had promised her sister in Detroit that she would call her as soon as her airplane landed in New York and she wasn't about to leave until she talked with her.

I explained the situation to her in German, but she wouldn't move. Finally, I assured her I would make the call for her, and stressed that she had to get on the plane now or she'd miss her flight and there wouldn't be another one for hours. With that she turned to me and said, "You know the problem with you Americans is you only speak one language."

What could I say to that?

The rattle of the keys unlocking the chain anchoring my leg brought me back from my reminiscing. The guard guided me back up the stairs and, as before, I was blindfolded and loaded into the back of a motor scooter and returned to my cell at the power plant. I never knew exactly what time of day it was, but it was often dark by the time I returned.

After my evening meal, I was left to stare at the four walls until bedtime. The next day the guard arrived at my door to take me to the underground factory once again where I went through the same drill as the day before. This routine went on for a week until I was relocated to another prison.

CHAPTER 6

Plantation Warehouse 13

THE PLANTATION, as the American prisoners called it, was located on the outskirts of Hanoi across from the Ministry of Defense building. The compound, formerly the home of the colonial mayor of Hanoi, and later the family residence of a French military officer during the French era, was taken over by the People's Army of Vietnam after the French left the country.

Once a beautifully landscaped villa with two acres of tree-lined streets, the Plantation contained a grand plantation-style house, called the Big House, a Show Room, Warehouse, Corn Crib, Gun Shed, and Movie House. Although the buildings stood in various stages of disrepair they made a suitable setting for displaying American prisoners to reporters, antiwar delegations, and dignitaries against a background of somewhat pleasant surroundings.

In late July 1967 I was transferred to a new location called the Plantation and placed in a ten-by-ten-foot cell in the Warehouse, a long, low building with a metal roof and shuttered windows. Although not as dreary as Hoa Lo prison, the grounds were enclosed by a brick wall topped with broken glass. The pale yellow-beige walls of my cell, number 13, showed scars from many years of neglect. A wooden door centered in the back wall partitioned my cell from another unused room behind which, I soon learned, spiders found a comfy home.

The bamboo ceiling, covered with tattered cloth, allowed guards to walk along the enclosed roof at night and spy on prisoners. A bare light bulb that burned twenty-four hours a day hung from the ceiling.

My bed was made of bamboo slats laid across two sawhorses. Two twelve-inch-long pieces of wood about a half-inch thick sticking out from the wall with a board across served as a shelf for my paltry possessions: a tin cup, a multicolored towel, clay water jug, toothbrush and paste, a bar of lye soap, and a second set of underwear I had been given while in the Golden Nugget. To show their "humane and lenient" treatment, they gave me my very own personal air conditioning unit (a bamboo fan) that failed to move the hot, sticky air. For bedding I had one blanket—no pillow—a bamboo mat, and a mosquito net.

Hello Kevin

As I slowly recovered from my wounds, I began hobbling around my cell trying to regain the strength lost during weeks of isolation and starvation. I must have shuffled around that tiny cell eight miles a day.

When the long, lonely hours turned into days, then weeks, I realized the world's greatest fighter pilot, was now little more than a caged animal.

One afternoon, about six weeks later, I was lying on my bed not doing anything in particular when I heard the squeak of the guard's rubber sandals approaching my door followed by the jangle of keys. Hearing the sound of keys sent a cold chill up my spine. It meant only one thing—I was going to be taken to an interrogation room for another round of questioning after which I would most likely be beaten again. What infraction of the camp rules had I broken this time?

I listened as the key was inserted in the lock and watched as the door swung open. There beside the guard stood Kevin, holding his bedroll, looking like something the cat wouldn't bother bringing home. My first reaction must have been shock, for I was certain Kevin had been killed that first night we arrived. He had aged from the last time I had seen him. His shoulders drooped and his face was pallid with dark circles beneath his eyes. Did I look as bad?

Before closing and locking the door, the guard announced that Kevin was being placed in my cell to take care of me.

God, it was good to see Kevin again. I couldn't have been happier if someone had handed me a million bucks. We stood expressionless until the guard's footsteps faded. Then we couldn't talk fast enough.

"How are you?"

"Okay, considering."

"What did they do to you?"

"Beat the crap out of me."

"Are you okay?"

"Does it look like I'm okay?'

Questions flew back and forth without waiting for the other to answer. Finally, taking a breath, I said, "Man, am I glad to see you."

"You look like hell," he said, and threw his bedroll down and gave me a big hug.

"Can't say you look any better."

After another round of back pats and hugs, we took stock of our situation.

"This is one big mess we've gotten ourselves into."

"Guess I broke the airplane."

"Yeah, they'll probably dock your pay."

Kevin might have gone through hell, but he still had a wit about him and that old McManus twinkle in his Irish eyes.

After our excitement died down, we began exchanging notes. He, too, had been subjected to hours of interrogation, indoctrination, and torture.

Kevin also had broken his back during the ejection from our airplane, but because he was able to walk a little better than I could, he chose to bring me my food and empty my honey bucket. But as I watched him limping around the room, I began wondering if he was going to take care of me, or if I would have to take care of him. It seemed Kevin's back was in worse shape than mine. His wounds just didn't show.

As part of my medical treatment, the Vietnamese gave Kevin a spoonful of sulfa to rub into my wounds and strips of dirty rags to use as a bandage. That was all the medical care I was afforded. But what I lacked in physical treatment Kevin made up for in emotional support. That was the best medicine I could have received.

It was wonderful having Kevin with me in this hellhole. We had always gotten along well at the Air Force Academy even though we didn't match intellectually. He graduated at the top of our class while I did no better than the upper quarter of the second half of the 498 new second lieutenants.

Having opted for a later flight training class, Kevin arrived in England two months after me. With his training complete, we caught up with each other at Bentwaters. By that time, I had already refused to fly with Major Incompetent and Kevin was next in line to be assigned to him. After flying only a couple sorties, Kevin came to me and asked why I no longer crewed with the guy. Having the same reservations as I had, Kevin went to the operations officer and also filed a complaint refusing to fly with him. Major Incompetent was relieved of his position with the excuse that the squadron was being realigned and crews reassigned.

With J.J. Davis assigned to a new aircraft commander, I needed a backseater, and since Kevin was available, and thinking we'd make a good team, I offered him the position. Kevin could have refused and picked an older, more experienced pilot, but he agreed to fly with me, making us the only 0-2 crew in USAFE responsible for flying a 3.9 million dollar aircraft with the mission to strike Communist Eastern block targets using nuclear weapons.

Kevin and I thought we were two hotshot pilots, while in truth, we were the most inexperienced pair of first lieutenants in a fighter pilot world filled with legends who had hundreds of hours of single-seat flying time and thousands of Command Barstool Association credentials.

We took every opportunity to fly cross-country to gain proficiency and experience while becoming familiar flying within international airspace. In other words, boondoggles. We would fly to Aalborg, Denmark, on Sunday for a fantastic breakfast buffet. On the way back we'd make a side trip to Spain and pick up watermelon, cantaloupe, and peanuts in Madrid, which Kevin carried on his lap.

While Jerri and I enjoyed living in the English countryside, Kevin, who was single, lived in a fully equipped BOQ (Bachelor Officers Quarters) with modern kitchen appliances. He had no difficulty finding dates and considered himself quite the lady's man. What he didn't realize was the English women weren't interested in him as an officer or pilot—they wanted his kitchen. Poor guy; so unsuspecting.

In 1965 the main focus was on the growing threat of the Soviet Union, but by 1966 the war in Vietnam had escalated from advisors to bombers with the Navy and Air Force pounding the Ho Chi Minh Trail the North Vietnamese used as a supply route to the South.

Believing the situation in Vietnam, considered just a police action at the time, would be over soon, we knew that in order to accumulate combat time, earn a row of ribbons, and be somebody in the Air Force, we had to get to Vietnam before the war ended. So we "volunteered" and received orders for Southeast Asia.

Again, Jerri packed our household goods and we said goodbye to our cozy little English cottage and got on the bus that took us to the airport. After getting Jerri settled into an apartment in Dayton, I headed to MacDill Air Force Base in Florida for specialized training in firing live radar guided missiles.

That's where I almost shot myself down.

Kevin and I were flying an F-4 carrying the AIM-7 Sparrow missile, which when fired, was supposed to lock onto a drone target. But as I fired the missile, it broke lock, lost its quarry, and seemingly locked onto our aircraft. The damn thing began following us, gyrating up and down looking for a place to do serious damage. When it looked as though the missile could actually strike our airplane, the safety observer yelled, "Cob it," which meant get the hell out of there. By shoving the throttles all the way forward, pushing the stick hard to the left, and lowering the nose to gain airspeed, I managed to break away, but the missile kept bobbing up and down until it finally plummeted into the ocean. I can testify that it wasn't fun having a live missile on my ass, and I believe we both had a slightly larger laundry bill that day.

But we still weren't finished with our training; we had two more stops to make before heading into combat. One was at Hamilton Air Force Base in California where Kevin and I spent a couple days mastering the M-16 rifle and another week in the Philippines, surviving in the wild jungle environment for three days. The training required us to travel several miles in a minimum amount of time as "the enemy" hunted us down. Getting caught, which we always were, meant we needed more training at evasion skills. Being clever fighter pilots, used to the comforts of the world, we learned from previous "campers" that we could avoid the ignominy of being caught sitting in the mud and, therefore, deemed in need of more training, if we, the "evader," gave the native Negritos a pack of Paxton cigarettes. By doing this, the natives saw to it that we didn't suffer the rigors of the harsh mountain terrain until we were "found" and returned to the base.

Combat in Vietnam

With orders for Vietnam in December of 1966, I was ready to put my training into actual practice. Interestingly enough, when the sixty-six pilots from Bentwaters got orders for Vietnam, everyone was assigned to the 390th Tactical Fighter Squadron, 366th Tactical Fighter Wing at Da Nang Air Base; that is everyone except me. There had to be a mistake and although I brought this error to the attention of personnel, I was told to wait until I got to Vietnam to correct the problem. When I arrived at Ton Son Nhut Air Base, however, a personnel weenie told me they had orders sending me to Phan Rang to fly the F-100 Super Sabre and the orders couldn't be changed. Wait a minute! Fly the F-100? That didn't add up. I was an F-4C pilot and an aircraft commander and here they were trying to send me off to fly an airplane I wasn't familiar with.

I tried explaining the situation, but the assignment officer wouldn't give in and kept insisting that I had to go to Phan Rang. I wasn't buying this deal. I wanted to be with the men I had trained with. I said screw it and hopped on the first C-130 going to Da Nang. There, I went to see the director of operations and reported with a salute and said, "I am an F-4C aircraft commander and have orders to fly F-100s at Phan Rang. I know zilch about the F-100. I'm here, can you use me?" He studied me a second, looked at my orders and then replied, "Okay, you're in the 390th Tactical Fighter Squadron."

After processing into the 390th, and before I could fly combat missions, I was assigned an experienced backseater to show me the ropes. Once familiar with the territory, rules of engagement, and operational procedures, I was put on the schedule to fly daytime and nighttime missions.

With my in-country training complete, Lieutenant Colonel Hoyt "Sandy" Vandenberg, asked who I wanted as a backseater. I didn't have to think twice, "If he'll fly with me, I'd like to have Kevin McManus."

Once again Kevin agreed, and we began flying together where fortune, fate, and a little luck set the course for Kevin's and my future as a combat crew. That choice saved my life for which I will always be grateful.

Our squadron's mission was to provide air-to-air MiG combat patrol for the F-105 Thunderchiefs. We would take off, fly north, link up with a KC-135, refuel in the tanker tracks over the South China Sea, complete the mission, egress, and then top off with another two thousand pounds of fuel before returning to Da Nang. Those were five- to six-hour missions usually covering 350 miles.

After flying escort missions into the Hanoi area, my squadron rotated to shorter nighttime missions of one to two hours that were flown in the southern Route Packages I and II. Although my daily routine consisted of regular bomb runs, there were times when I was placed on standby alert for close air support in the South. When flying these missions, we provided protection for the defoliant spraying C-123 Fairchild Provider by laying down a corridor of CBUs (cluster bomb units) and smoke canisters. Although the C-123 with its two piston engines augmented by two small jet engines was a great aircraft, it was slow, making it a perfect bull's-eye for bullets and missiles.

As a new aircraft commander, my main concern when flying my first close-air support combat mission with live bombs was not screwing up. In order to release my ordnance, I had to push the weapons select button on the dog bone by flipping three switches, but when I did, nothing happened. I figured I must have done something wrong, and began sweating bullets knowing I had blown it, but after running the usual post flight check, the munitions crew found a short in the system. I breathed a sigh of relief; I hadn't screwed up after all.

I never gave much thought about pitching ammunition from my aircraft that could actually kill people. Most of the time I was too busy flying the airplane and dodging antiaircraft fire to have time to look and see the impact my bombs had made. I saw buildings and trucks blow up, but never stopped to consider the people inside. When you're flying at 600 knots, everything happens in nanoseconds. After dropping my bombs, all I had time to do was to pull up and get out of range of the big guns firing back at me.

But there was one time after dropping a bomb at a low altitude that I did turn and look down at my hit. Maybe I shouldn't have because I saw people die that day. As a pilot you learn not to personalize your actions; all I could feel was, as General William Tecumseh Sherman once said, "War is hell." He was right. You disassociate yourself and get back to doing your job.

On January 10 during our first daytime deep penetration mission into North Vietnam we pulled up behind the tanker and I said to Kevin, "I've never done this before from the front seat." If Kevin had reservations about flying with me at that moment, he never said anything. I looked around the cockpit and found the button for the refueling receptacle, pulled the knob, and held my breath. Because the refueling boom went into the airplane behind the cockpit, it took both of us, each working a different part of the controls, to accomplish the maneuver. I directed the stick back and forth, left and right, while Kevin, ran the throttles. This took coordination, a steady hand, and trust. That trust later carried us through some pretty rough times.

Kevin and I made a good team. What I didn't know, he did. Where he lacked training, I took up the slack. We worked out a system where he knew when I would fly the airplane and I knew when he should take over. That was fortunate for me, because he made me look good on more than one occasion.

Fighter pilots are resilient. We bounce back and go on, and sometimes we have been known to party rather exuberantly. Hey, we worked hard. It was war out there. We deserved some happy time. Since there was very little to do during off duty hours—no organized sports or golf courses—we were left to lifting weights and studying maps, and from time to time we downed a few at the Doom Club, Da Nang's officer's open mess.

It was on one of those nights that we crawled into our bunks around midnight after imbibing at the club when we got a call at 0300 to launch on an in-country mission to the mountainous region on the Laotian border about a hundred miles southwest of Da Nang to help support troops that had become pinned down by enemy fire coming from the mouth of a cave.

Once over our target, we surveyed the geography. We were briefed on the enemy's location and told who was who to avoid hitting the good guys. To strike a mountainside, the SOP states that if you use a forty-five degree dive, you'll be too steep. If you use a slightly shallower angle it's easier to throw bombs into the side of the hill.

"Okay, Kevin, give me a thirty degree dive bomb setting," I said.

I cranked the depression mils into the gun sight. We rolled in and dropped the bomb right into the mouth of the cave. Because of the elbow bending the night before, I wasn't in the best condition shall we say, and as we pulled up my blood pressure dropped and everything started to go dim.

"Kevin, take the airplane. I can't see."

He pulled the airplane up and started laughing.

"What are you laughing at?"

"I gave you a thirty degree dive bomb setting, right?"

"Yep."

"You had the airplane at a forty-five degree dive and the bomb went right into the cave."

"Okay, don't change anything. Let's go around again," I said.

We went around three more times using the thirty-degree setting with a forty-five degree dive. On each go around, I took a trip to la la land and Kevin pulled us back to a downwind position and we'd roll back in. We dropped six good bombs and the forward air control went nuts because we had blasted the cave into extinction.

After that, I made it a rule never to fly hungover again.

Another night we had just dropped our bombs and the Vietnamese were firing at us with everything they had. I made a hard right turn away from the tracers of the 50-caliber bullets streaming toward me when all of a sudden the airplane veered off in the opposite direction.

"What happened?" I yelled to Kevin.

"The 37mm guns were bigger on this side," he said.

A couple weeks later, Kevin and I were flying our first escort mission into Hanoi, positioned on the right wing of Bill Baugh and Don Spoon in a standard four-ship formation with our jamming pods on. Bill and Don were in the number three position on the right at fifteen hundred feet out. Kevin and I were slightly back and five hundred feet above Bill and Don.

The plan was to create a broad jamming strobe back to the SA 2 Fan Song guidance radar so the Vietnamese couldn't achieve a "lock" on our aircraft and fire at us in the fully guided mode. Despite our tactical formation, they spotted us and fired. Bill and Don's airplane took a hit in its gut by a large caliber antiaircraft shell. I watched as their airplane tumbled out of the flak-filled sky and burst into a glowing orange fireball as it hit the ground. Not seeing a full parachute, I had no way of knowing if they ejected safely.

I should have felt angry, maybe even sad, but when the bad guys are shooting at you and flak is exploding all around and pilots are yelling over their radios, everything is pure bedlam. All I could do was watch the radar warning signals light up like a Christmas tree and hope I wouldn't get hit. I didn't have time to think about losing fellow pilots. All I could do was offer a prayer, and then hightail it out of there.

Being in Vietnam where the guns were real and the bullets could actually punch holes in my aircraft gave me a new respect for the word combat. It was then that I truly became aware that flying in enemy territory was a deadly business.

A fighter pilot never thinks he'll be the one knocked out of the sky; it's always going to be the other guy. But the other guys were men I had flown with in Europe. I knew Bill's family back in Piqua, Ohio, and seeing him get shot down was personal, but I couldn't dwell on it. I had a job to do and knew I would get up the next day, tempt fate, and do the same thing all over again.

The flight before Kevin and I were shot down was a precursor of things to come. After dropping our bombs, we had somehow gotten down to a two-ship formation and were separated from the rest of the flight below the clouds. At that relatively low altitude, the sky was soupy and visibility was nil. I didn't want to be beneath the clouds where I was in danger of getting hit, and being in the clouds, I couldn't see the SAMs coming. It was a no-win situation.

There we were, boogying around at a low attitude, trying to find our way out of the mess when Kevin told me we had drifted over an area we weren't supposed to be. Within seconds we took a hit. Climbing rapidly above the clouds, I looked back and saw a flickering fire in the right engine. I shut it down and watched the fire go out. All was well, but the airplane was low on fuel.

Knowing we couldn't make it back to Da Nang, I headed for Udorn, Thailand, and called for fuel, pleading actually, for a tanker to come to us because we certainly couldn't get to them. A tanker from U-Tapao Air Base came out of its normal, and somewhat benign, tanker track to lend us a hand. As I maneuvered into position to take on fuel, the boom operator asked if I knew I was on fire. What! I knew I had shut down the engine and saw the firelight go out. What I didn't know was the fire had burned through the circuitry board causing the light to go out but the airplane's right engine still blazed.

Flying at twenty thousand feet with a ruptured fuel line, an engine on fire with no way to extinguish the flames, and a KC-135 filled with thousands of gallons of highly explosive JP-4 jet fuel was not an ideal situation to say the least. I expected the KC-135's aircraft commander to say, "Sorry, fellows. See you later, and, oh by the way, good luck with your problem." But he hung out his cajones and plugged into us.

As the tanker pumped fuel into my airplane, the left engine sucked it up, but the right tank kept dumping fuel out as fast as it went in. There was no way with the engine losing fuel that we could make it to Thailand, safely. Thankfully, the tanker stayed plugged into us all the way to the air base.

Fighter pilots have huge egos and make jokes about anyone who doesn't fly a fighter, but I have to hand it to that tanker crew. They pulled my fat out of the fire that day and I will forever be indebted to them. I made it a priority to find that crew and thank them personally, but I got shot down on my next mission and never got the chance. I will say this, they were one hell of a courageous crew and I've never said anything bad about trash haulers or gas passers since.

At Udorn, the maintenance crew fixed the engine while Kevin and I hitched a ride back to Da Nang on an old C-47 Courier. Several days later a maintenance crew flew our F-4 back to the base. On the airplane's next mission, I just happened to be the unlucky bastard assigned to fly that same airplane again. I don't know if my unceremonious departure from that aircraft was the result of being hit by a missile or if the repaired engine blew.

We'll never know.

By early June 1967 I had already flown 113 missions with 79 of those over North Vietnam and was nearing rotation. I was pumped about my next assignment to Duluth, Minnesota, where I would be serving with the Air Defense Command protecting our northern borders from Soviet bomber attack. What made the transfer so exciting was that I would be assigned to fly the F-106 Delta Dart, an airplane I had dreamed of flying since seeing a very impressive aerial demonstration of it at Vance Air Force Base.

There was an unwritten rule that if you had flown more than seventy-five missions over North Vietnam you didn't have to fly the dangerous Route Package VI missions over Hanoi. My last twenty-five missions were supposed to be milk runs, flying the lower, safer Route Packages I or II, but because the squadron didn't have enough guys who had flown the required ten "get acquainted" missions before being assigned to fly into Hanoi, and because a lot of men were finishing up their tours and going home, we were short of pilots.

So, on June 14, 1967, when Kevin and I were asked if we would mind going to Hanoi, even though this would be my eightieth mission, I said what the heck? What was one more mission?

That fateful day began like all the others with an intelligence briefing around 0900, for a 1400 time over target. Our assignment was to hook up with F-105s, air-to-ground bombers, coming out of Takhli and Korat, Thailand, in the tanker track over the South China Sea. Once given our ingress altitude, air speed, radio frequencies, locations of the SAMs and gun emplacements, and egress information, we were ready to fire up our machines and hit the sky.

On the flight line, while Kevin made the weapons check, I conducted the preflight, checking gauges, fluids, and kicking the tires. Satisfied that everything was in order, I climbed the ladder to the cockpit and strapped in. After running the flight control check, I returned the salute of my crew chief and we taxied for takeoff.

Flying under the code name Rolling Thunder, and using the call sign, Chisel 04, I was part of a fifty-four aircraft strike mission to hit the major Vu Chu railroad yard and rail lines that ran out of Hanoi up into China. Normally, the F-4s were placed in a bomb stream of two flights of Phantoms, four abreast, inserted into the third position of the stream to shield the F-105 strike force from Vietnamese MiG-17 and MiG-21 jet fighters.

But this mission was different. Conventionally, F-4s flew MiG CAP to chase MiGs if they showed up, but since we hadn't seen any MiGs in some time the intelligence folks from 7th Air Force decided to make us bombers for the day.

My airplane was last in the four-ship formation. After refueling over the South China Sea, and staying underneath the fifteen-thousand-foot cloud layer, we flew inbound to the target area between Hai Phong and the Chinese border, heading toward the southern most entrance of the railroad yard. That's where I dropped our bombs, got hit, and ended up as a guest of the North Vietnamese.

UFOs

Kevin and I had a lot of catching up to do. He told me that at one point he had six missiles on the radar-warning screen in addition to the hundreds of flak explosions coming from the 37, 57, and 85mm antiaircraft shells popping around us. Not that it mattered much. We still ended up ass over teakettle as we ejected.

He related the story about the Vietnamese fighter pilot who came to visit him twice while at the Power Plant. (I didn't even know he had been at the Power Plant.) The pilot, wearing a clean and neat purple flight suit, would come in, sit down and look at Kevin. He never spoke, and after a few minutes, got up and left. Kevin said he was not aggressive, never displayed any animosity; he just sat there as if visiting an old friend. Kevin never found out what it was all about.

After swapping notes on our current situation, we settled in and began reliving our good, old pre-shoot down days at Da Nang like the time we saw a strange object in the sky.

We had taken off from Da Nang and were at an altitude of eighteen thousand feet, two miles north of the DMZ heading south when Kevin got a huge return on the radar. I had a radarscope repeater in the front seat and could also see the blip. It was two to three times bigger than any return I had ever seen coming off a conventional aircraft.

Kevin locked onto it. Ordinarily, if you're closing in on another airplane, you have your airspeed of three to four hundred knots and the other aircraft's airspeed of three or four hundred knots. Combined you're looking at 600 or 700 knots cruising speed. But our range gate indicated the closure speed of this object was 2700 knots and closing in on a collision course about fifty to sixty miles out. What the hell kind of new weapon was this?

Within seconds I spotted a large purplish, almost black blob in the eleven o'clock position, coming at us. I lit the afterburners and shoved the stick forward to avoid running into it. I know it passed us, but when I turned to look, it had disappeared.

After landing at the air base, I got on the classified line to Saigon and reported what we had seen. It took three days of telecoms with Kevin and me telling the same story before some colonel called me and said, "I don't know what you saw, and we don't know what it was, but don't talk about it any more." That certainly cleared everything up.

There were two other times when Kevin and I saw strange objects in the sky. This was one of those "dark and stormy night" tales. I was in the radar pattern being vectored around to land at Da Nang when all of a sudden the radar ground controller said, "Chisel 04, you have a bogey at your seven o'clock, one mile out."

The radar controller gave me a turn to get out of its way, but the bogey turned with me. He gave me a series of six turns and each time the bogey stayed on my tail. Now low on fuel, I had to land. The controller lined me up, and as I came in on final approach the object disappeared. I looked all around. It had vanished.

One other time, while sitting on alert the crew kept seeing a pulsating purple light sitting off the end of the runway. Airplanes were taking off and landing and the light continued sitting there blinking for hours, never moving. The F-102s used for air defense alert were scrambled to see what it was. When they took off, so did the light. They pursued the light at 500 knots for 250 miles until the strange light finally disappeared out over the South China Sea.

A couple nights later we saw the light again at the end of the runway. The light sat out there all night as if watching our airplanes taking off and landing, but when our F-4s were scrambled to find and identify it, as before, it took off. It was as though it knew we had been specifically tasked to discover what it was. We followed it for five hundred miles out to sea while it played with us, zigzagging through the night. We kept it on radar, but could never catch up to it. Then, like before, it instantly disappeared.

Was what we saw a UFO? Who knows? But if I ever had doubts about strange objects in the sky, I'm a true believer now.

Our days at the Plantation never varied. We awoke each morning around 0600 to the sound of a clanging gong, dressed and emptied our honey bucket in a slough beside the bathhouse near the Corn Crib. Then it was back to our cell for more wall staring, interrogations, and attitude adjustments until 1000 hours when we got our first meal of the day—pumpkin soup, a piece of stale bread, sometimes a small lump of rice, and on very rare occasions a banana.

After breakfast, and I use that word laughingly, a guard came around and gave us our daily ration of three cigarettes and lit one from a punk stick that was nothing more than a smoldering piece of wood with a glowing tip used as tender. Because the guard always watched to make sure I smoked the cigarette, I began puffing away at what tasted like dried roach droppings. I had never developed the smoking habit, but Kevin was a chain smoker. Once the guard closed the judas door, I'd put out my cigarette and give it to Kevin so he could smoke it later.

Then we sat counting the bugs on the wall until 1400 hours when we got our second meal of the day—pumpkin soup, a piece of stale bread, and a lump of rice. The menu never varied. The afternoon meal was followed by another cigarette. Sometimes we were given chores to do and sometimes we were subjected to Vietnamese propaganda in the hopes that we would come to understand the error of our ways.

When the afternoon gong sounded for quiet time, we were ordered to sit silently on our beds. No talking. No lying down. We never did for fear a guard might peek in through the judas door and see us violating camp policy, which would have meant a beating. We sat like that until the gong sounded again indicating siesta time was over.

We followed that same routine, day in and day out, for several weeks until August when Kevin and I were told to gather our belongings; we were being moved.

CHAPTER 7

Plantation Warehouse 6

IN AUGUST KEVIN AND I WERE TAKEN from Warehouse 13 and placed in Warehouse 6 where we met Joe Milligan, a pudgy, balding Rutgers University graduate, born and raised on a dairy farm in New Jersey, and Gary Anderson, a short, stocky weight lifter from California. I had gone from isolation,to having one cellmate, now I had three. Four men living in one small cell proved interesting.

The room in Warehouse 6 was slightly larger than the one Kevin and I had previously occupied and had small slits in the boarded-over windows that let in tiny beams of light and a little air—very little air. Although the walls were dingy and yellowed, the decorative tiled floor gave faint evidence that the room had once had a more pleasant use.

Communicating

Communicating, using the Tap Code, by tapping on the wall to the man in the next cell became our main means of sending and receiving messages. Even though there were times we tapped until our knuckles bled, we continued. It was paramount to our sanity and worth the risk of any punishment the guards could dish out if we were caught.

The Vietnamese knew we had a method of communicating and tried to catch us in the act by tapping on the walls themselves to get us to respond. One afternoon a guard came into my cell, called me by my Vietnamese name, Bia, which means beer by the way, and told me to tap on the wall.

"I don't do that. Tapping on the wall is against camp regulations," I told him.

"Tap on wall," he ordered.

When I refused, he yanked me outside for a little physical persuasion.

"Tap on wall," he said with renewed vigor.

To avoid another pummeling, I tapped gently on the wall a couple times.

There was no response, of course.

"Tap louder," he demanded.

So, I tapped louder. Still, there was no response.

"Louder!" he shouted.

Boom! Boom! I pounded on the wall. Still nothing.

It seems they were trying to con Jim Kasler, who lived on the other side, into replying, which would prove we had been communicating.

Not getting a response, they went into Kasler's cell. The walls were thick, but with the doors open I could hear them ask, "You hear tapping? Why you no answer?"

"What tapping?" Kasler asked.

The Vietnamese hadn't figured out that we always began each message with the old "shave and a hair cut" signal. No shave and a hair cut meant the message wasn't coming from an American. We also had a warning signal of two hard thumps on the wall if a guard approached while we were "talking."

Along with the tap code, men invented several ingenious methods to communicate. Because respiratory problems were common, the TB (Tuberculosis) code became a convenient and covert method of sending messages. By coughing, hacking, snorting, wheezing, sneezing, sniffing, grunting, spitting we could "talk." If we could have found a way to pass gas on cue, I'm sure someone would have used that method, too. Watching men hack and cough messages across the courtyard was hilarious. Had I tried using the TB code, I would have cracked up laughing.

We also found that by holding our tin cup to the wall and talking into it we could amplify our voice and communicate with the man in the next cell. Our voices sounded muffled, but it worked.

Others devised a hand signal using the alphabet, much like the American Sign Language. The "Q" had a particular significance—the fighter pilot's salute. The use of hand signals, however, was limited to a direct line of sight as one man signaled through the vent hole near the ceiling to another prisoner in a nearby building.

Brooms in Vietnam, unlike the soft-bristled brooms we have in the United States, were made from hard sticks of bamboo fastened with a hemp-like cord, making its use ideal for the Sweeping Code when cleaning the dirt grounds of the compound.

Late one afternoon while sitting in my cell I heard the familiar sound of swish, swish, pause, swish, swish. Someone in the courtyard was sending a message. Never knowing when something important was coming across the Comm Net, I tuned in to listen.

The message: "Mary Poppins was a junkie."

Everyone within hearing range burst out laughing, which threw the guards into a frenzy, running around yelling, "Shience! No laugh. Shience." They had no sense of humor.

Pig Duty

That December Kevin and I were assigned the wonderful job of cleaning the pigsty adjacent to the Big House. For some guys, I suppose, cleaning a pig's pen would seem demeaning, but for us pigsty duty was a welcomed change from sitting in a stinking, cramped cell all day with nothing to do. Even though it was the middle of winter and quite cold, the work gave us a chance to see sunlight for more than the few minutes a day we were given to empty our pots and bathe.

Along with our pigsty cleaning duty, Kevin and I also had the privilege of feeding the smelly beasts. That's when I discovered the guards weren't the only ones with a bad disposition. The pigs weren't all that friendly either.

Despite the unpleasant smell, we took our time feeding the pigs, milking the job for all it was worth, while carrying buckets of warm corncobs and slop from the kitchen to the pigsty close to our bodies. My stomach rumbled as I smelled the aroma of warm corn. The damn animals ate better than we did.

Over time the pigs became accustomed to our presence and allowed us to get close enough to feed them by hand. Being a fearless fighter pilot, I took a handful of chopped cornhusks and held it for a pig, expecting it to lap the grub like a dog. However, I quickly learned the little devil had sharp teeth when it clamped down on my fingers. I thought that critter was going to eat my entire hand before letting go. I wore its teeth marks like a badge of honor for weeks. Probably should have gotten a Purple Heart now that I think about it.

One of the piglets wasn't doing too well and had been taken to the kitchen to keep warm until it grew stronger. When it was up to par, I was told to take the piglet back to its pen. As I walked across the courtyard, carrying the piglet in my arm, the camp commander happened to come strolling toward me. Being required to bow to all Vietnamese to show "good attitude," I bowed. Having no other choice, the pig bowed too, but instead of being grateful that it wasn't being served for dinner, it let out a squeal as if being slaughtered. The guard standing nearby started screaming, "Don't bow to commander with pig." Bow. Don't bow. What was I to do?

Taking the Easy Way Out

Being held prisoner didn't stop us from continuing to serve our country. Although our uniforms and duty assignments were different, we still had a job to do.

The Warehouse sat at one end of the compound, with three similar buildings occupying the other end of the camp. Two buildings sat empty in the middle, dividing the camp into East and West with Navy Lieutenant Commander Dick Stratton at one end and Air Force Lieutenant Colonel Hervey Stockman at the other. As the senior ranking officers (SROs) in their area, both men took charge of their end of the camp.

Communicating was our lifeline, and for some, a means of survival, but the divided camp made it difficult to send messages between the two SROs. Cleaning up after the pigs gave Kevin and me the perfect opportunity to pass along information.

At the direction of the senior officers we were responsible for relaying information to individuals isolated in solitary. The purpose was to make sure their experience was not a personal one-on-one situation with them against the Vietnamese. However, there were cases where a man was cut off from the group because we couldn't get to him, and in a couple cases, the men didn't give a rat's ass and wanted to be left alone, never becoming part of the community.

When we got wind that a couple men were side-stepping the Code of Conduct and violating Article III which states: *I will accept neither parole nor special favors from the enemy,* and were cooperating with the Vietnamese in return for an early ticket home, it became our duty to convince them to reassess their actions.

David Matheny was one of the men we suspected of taking the easy way out. His cell was about ten feet from the pigsty, making it relatively easy for Kevin and me to relay messages to him from the SROs. While Kevin and I cleaned the pig's pen we pretended to sing like happy farmers. It wasn't any particular melody, we would hum under our breath just loud enough for him to hear, "Hey, Dave, the boss says don't go home. Don't let the Vietnamese put pressure on you. Colonel Stockman wants you to understand, don't give in. Don't do it. Hang in there."

We cleaned and sang everyday, "Remember, no special favors. Hang with us. We're all in this together. La la la."

Kevin and I knew Matheny and two other prisoners were receiving special treatment, because we observed them receiving meat and more than one banana a day. They were also being fed three times a day and on nicer plates than the rest of us.

When Matheny ratted on us to the camp commander, saying we were trying to communicate with him, Kevin and I were hauled into the camp commander's office. I expected we'd be beaten, but the commander used the opportunity to tell us that if we showed "good attitude" and admitted the wrong we had done and came to appreciate the humane and lenient policy of the Vietnamese, we, too, would receive special favors; we could be released early and reunited with our families. Of course, we blew him off. But instead of being punished, we got "fired" from our prestigious pig duty job. We were devastated.

Despite all our efforts, Jon Black, David Matheny, and Norris Overly were released in February 1968 to Father Daniel Berrigan, a Jesuit priest, and Howard Zinn, a history and government professor at Boston University, both members of the National Mobilization Committee to End the War in Vietnam.

The effect of the early release rippled through the camp, leaving us feeling betrayed. They were leaving under less than honorable circumstances after willingly aiding the enemy. Where was their sense of duty and honor as officers? Where was their integrity and self-respect?

It would have been easy for all of us to sell out and accept special treatment or favors so we could go home. Who would know if we took a cup of milk or extra banana? What difference would it make if we made an antiwar tape and spoke with a delegation?

Instead we chose to take the moral high road, and though that decision resulted in a great deal of physical discomfort for many, we hung tight. We were going home with honor and our heads held high. Even with our moral fiber tested time and time again, we had the courage to stay true to the Code. We can live with our decision. I wonder if they can?

The division of the camp also made for an awkward situation when Seaman Doug Hegdahl got caught between the two SROs: Stratton, who was Hegdahl's roommate, told him that since he was enlisted and had not been captured like the rest of us, but by accidently falling overboard from his ship, if offered an early release he should take it. On the other side of the camp, Stockman gave an order that under no circumstance was anyone to take an early release, including Hegdahl. That placed Hegdahl between a rock and a hard place.

Stockman's order for Hegdahl to stand firm was countermanded by Stratton who gave a direct verbal order for Hegdahl to go home and said that he, Stratton, would accept responsibility for that directive. His justification was Hegdahl had memorized the names of all the men being held captive and could report that information to authorities once home. However, we all had those names and were told to memorize and recite them to our cellmates three times a day just in case one of us did happen to get a sanctioned release.

In Hegdahl's defense, he didn't want to go home early. In my opinion, he's an honorable man compared to the SOBs who kissed ass and made nice-nice with the enemy. Unlike so-called peace advocates, Jane Fonda, Joan Baez, Stokely Carmichael, and Ramsay Clark, who claimed that we were not being tortured and, in fact, were being treated humanely, Stratton's decision to send Hegdahl home also gave him the opportunity to spill the beans and set the record straight about our confinement and abuse.

Just as a few men—and I emphasize a very few men—would stoop to anything to curry special favors from the guards, Tom Sumpter, who lived next door, stood at the opposite end of the spectrum.

Tom, a quiet, conscientious guy, took adherence to the Code seriously and became concerned when a guard, for no apparent reason, gave him an extra ration of four cigarettes instead of the customary three. He feared that the guard might be trying to butter him up to gain his confidence and get information from him.

Tom began tapping furiously on the wall, "Did anyone else get four cigarettes? Why did I get four cigarettes? Do you think they're up to something?

Are they trying to break me down?"

When the guard didn't request anything from him, he finally calmed down. For all we know, the guard could have made a mistake or could have been in a rare and generous mood and felt sorry for Tom because he had been isolated so long. We'll never know, but Tom sure caused an uproar for a couple days.

One night, I was awakened by tapping coming from Tom's cell. Thinking that something important was happening, I sat up, listening carefully to the message.

"Look, there's Venus."

What in the devil was that supposed to mean?

Being alone for so long, Tom had created his own form of entertainment by spending his nights peering through tiny holes in his boarded window, watching the stars. Being a navigator, Tom had knowledge foreign to mere fighter pilots. He knew the names of all the stars, and whenever a star appeared in the night sky he would start tapping to tell us what it was. It didn't matter to him that it was the middle of the night and had awakened us. And he seemed totally oblivious that guards were walking around and could hear him tapping; he just tapped away. Because of his midnight ramblings, I gained a lot of useless celestial knowledge. I now know where all the planets are in the spring and winter sky. Hoorah!

SIUTGA and SIUYOA

As much as we enjoyed thinking we had some control over our lives, it was the Vietnamese who came out on top in the end. We were often beaten if we didn't agree to meet with an antiwar delegation or write an antiwar statement applauding Ho Chi Minh or make a tape that would be played to soldiers in the South. But we stood our ground, took the beatings, and refused to obey their demands. Once the Vietnamese realized that we weren't going to give in and that physical persuasion wasn't going to work, they tried a more subtle approach of "non-physical inconveniences."

We found cunning ways to annoy the guards and invented the terms SIUTGA and SIUYOA to identify our deviousness. SIUTGA, an abbreviation for Stick It Up The Goons Ass was used when we got one over on the Vietnamese. SIUYOA meant Stick It Up Your Own Ass—that boiled down to our cleverness coming back to bite us in the butt. Whenever we rebuffed a guard's overture or refused to do something the Vietnamese wanted, what came back to us was no cigarettes or bathing for a week—just a subtle message that told us they were still in charge.

Messing With the Guards

We loved sticking it to the guards and used every opportunity to irritate and confuse them. When the Vietnamese began putting numbers on our clothing to identify us when taken for interrogation, we found a way to mess with them.

After washing our clothes we hung them on the clothesline in the courtyard to dry. As we took our clothes off the line we'd swap them with another prisoner's. When a guard came to take us in for an interrogation session, he'd look for us by number, not by name.

Once, I was taken to the Big House for questioning and showed up wearing a different number other than TU513, the one I'd been assigned. The interrogator looked confused when he saw that the number on my uniform didn't match the name on his paper. I got a kick watching his expression. I'm sure he knew he had been duped, but to save face, he talked to me anyway.

When numbering our clothes didn't work they tried fingerprinting. I was again taken to the Big House where they had set up shop on a table with an ink-pad and a small square card. The guard held my index finger, rolled it over the ink, and then pressed the tip of my finger onto the paper. Always looking for an opportunity to monkey around with these halfwits, I decided to have some fun. When the guard rolled my finger, I pressed down hard, smudging the print into a solid blob of ink. The guard wiped the ink off my finger and repeated the process. Once again I pressed down hard. Getting the same results, he looked at me and said, "Your fingers, they are sick. You have no lines."

"Yes, I know," I said, choking back a chuckle.

"When will they get healthy?"

Did I say halfwit?

Word quickly spread through the camp and soon all the guys came down with a case of sick fingers. The guards never caught on to what we were doing.

SIUTGA

Coal Ball Duty

While Joe and Gary were off on an extra detail somewhere, Kevin and I were given the job of making coal balls by taking a mixture of sand, coal dust, mud, and water, and molding it into something similar to charcoal to use for cooking. At first we refused, not wanting to help the Vietnamese, but they convinced us that the coal balls were used to cook our food. After getting the okay from our SRO, Hervey Stockman, Kevin and I became coal ball makers. Actually, it wasn't so bad because the job allowed us access to the warm kitchen where we stacked the balls near the stove.

The balls had to conform to specifications: no smaller than a golf ball but not larger than a baseball. Making coal balls was a dirty job, and after making a huge stack of them we were covered with black dust. Gold Tooth Fairy, a guard aptly named because of his one front gold tooth and because he pranced lightly in his sandals, took Kevin and me behind the Warehouse to bathe in the cold-water cistern. There, he made us strip off our clothes and stand naked shivering in the cold as we washed our hands and splashed water over our bodies. Although most Vietnamese seemed uncomfortable by nudity, this guy seemed to enjoy the view. I did say Fairy, didn't I?

The cistern was located next to an elevated train track where passengers could look down and see us, but more importantly, they could see the important duty Gold Tooth had guarding the two naked American Air Pirates.

After we had washed sufficiently, he offered us a cigarette. I suspected his purpose was to keep us standing there naked until a train came by. As Kevin and I smoked, Gold Tooth lit up, too, and like a kid with his first cigarette, held it between his thumb and index finger and pretended to smoke by puffing without inhaling.

Kevin got *that* look in his eyes that told me he was up to something. Making certain that Gold Tooth was watching, Kevin inhaled by drawing in a long drag of smoke, blew it out, and then French inhaled the smoke back in through his nose.

"Ah, this is good," Kevin said, and carried on as if it were the best cigarette he had ever smoked. I copied Kevin, making smoking an art form of pure ecstasy as Gold Tooth watched, intrigued by the American way of smoking.

Then Kevin suggested that he try it. Eager to be like an American, Gold Tooth drew in a lungfull of smoke and immediately began coughing and hacking just as a train rolled by. We watched people on the train looked down at this Vietnamese guard barking and stomping around. Humiliated, he threw down the cigarette and marched us back to our cell.

We didn't dare laugh then, but we let loose once back in our cell. It felt good to put one over on him.

SIUTGA.

CBUs

Younger, less experienced guards got excited easily. One night we heard the thunder of American aircraft overhead followed by CBUs (cluster bomb units) used to suppress antiaircraft fire being dropped on the nearby Paul Doumer Bridge. Pellets from the CBUs were no bigger than buckshot but when striking the tin roof of our building sounded like a squad of tap-dancing raccoons.

One young guard, believing we were in danger of imminent death, flung open our cell door holding a handful of pellets yelling, "CBU! They kill you! CBU! They kill you!"

We looked at him, then at the killer pellets, and crawled back onto our beds.

A few days later, we heard bombs being dropped again on the Paul Doumer Bridge. The combination of bombs pounding the ground and the returned antiaircraft fire caused the earth to rock and roll like a Jell-O shot. Everything was heaving and shaking so severely that none of us in the cell could stand up. Bombs were making one hell of a racket, guards were running around screaming and yelling, magnifying the chaos, and all we could do was crawl under our bunks, clasp our arms over our heads, and wait for the attack to end.

Through all the uproar, I hear tapping. Thinking that someone may have been injured, I scooted close to the wall to listen.

From Jack Van Loan next door comes the message: "Polo anyone?"

Ya gotta love a guy with such a warped sense of humor.

Tooting Contest

As stifling hot as the weather was during summer, the winters were even worse with temperatures often falling below thirty degrees. Without much clothing and no heat in our cells, we spent many bitterly cold nights shivering under one thin blanket.

It was on one of those bitterly cold nights that Kevin and Gary got the brilliant idea to stay warm by sharing their blankets. The theory was: two blankets over two bodies was better than one blanket over one body. Plus the additional body heat would add to the formula and help keep them from freezing. Made sense. So, Gary and Kevin doubled up under their two blankets.

It didn't take long for the mischief to begin. It started with a tiny toot and a "Sorry. Excuse me."

That was followed by a slightly louder and longer toot. "My God, what did you eat?"

"Shut up and go to sleep."

Then, like two kids at summer camp, they got into a farting contest.

"Anyone have a gas mask? My eyes are burning!"

Luckily, they didn't have matches or their antics would have turned into a flatulent flame-throwing contest.

The experiment lasted exactly one night.

Vietnamese Toilet Paper

We could have gone without many necessities, but toilet paper wasn't one of them.

The Vietnamese had no concern for our sanitary needs, and we went long stretches doing without. When we did get toilet paper, which I can only describe as sandpaper embedded with twigs and straw, there wasn't much of it. The guards came around every once in a while with a two-by-three-foot sheet of paper and tore off a strip and gave us each a piece. It wasn't much and it didn't last long. In desperation, we were left to steal anything we could find lying around the compound suitable for the purpose. The guards were not pleased when they found pieces of newspaper with a picture of Ho Chi Minh in our toilet paper pile.

SIUTGA

Dishwashing Duty

Still using us as cheap labor, Kevin and I got KP duty, kitchen police in Army terms, which meant we were the official dishwashers. It wasn't a bad job and, like pig duty, got us out of our cells for a few extra minutes each day.

While scrapping leftover grub into a bucket for the pigs, we found messages scratched under the rim and bottom of the porcelain dishes. By using aluminum spoons, men discovered they could scratch messages on bowls and plates. Guards never sullied their hands with garbage and had no desire to inspect the dishes, making it easy to use them as a means to communicate.

Hervey Stockman, who was the first man to fly a U-2 spy plane over the Soviet Union, wrote messages on the back of the porcelain bowls and plates when he wanted to send guidance and directives to the rest of the men, such as stay strong, don't give into the enemy, and don't violate the Code of Conduct.

Having access to his messages, it was Kevin's and my job to see that his memos were passed along to the right person. Soon, we were like a well-oiled machine receiving and sending messages across camp between the SROs.

Aside from using dishes as a communication device, Stockman, a talented artist in his own right, drew beautiful pictures and cartoons on the back of the plates. Unfortunately, no one saw them except Kevin and me. It was amazing that he could create such wonderful art with only an aluminum spoon. Sadly, we had to scrub off all his drawings after each meal and they were lost forever. Such a shame; they were very special.

Are You Who You Say You Are?

"Swede" Larson, a new arrival, was put in isolation. Kevin and I were pressed into duty using our Winged Mercury Messenger Service to make contact. Employing brick dust, chips of wall peelings, or cigarette ashes mixed with a little water or spit, and sometimes blood, we made ink and wrote messages on pieces of scrap paper found lying around camp. We secreted Stockman's instructions everywhere we thought Larson might find them and where we knew guards wouldn't look, such as the wash stall of our bathing area and the sewer drain.

Larson, the doubting Thomas, wanted verification that Stockman was in fact who he claimed to be. He insisted on authentication that the information we were risking our lives to give him was legitimate and coming from an American and not the ploy of the Vietnamese setting a trap.

Stockman answered Larson's question only to have another question shot back asking for more proof. Messages sometimes took several days to reach its intended receiver and this jousting went on for weeks with messages being carried across camp.

One of Larson's questions was, who were the members of the aerobatic team of the Star Dusters at Luke Air Force Base in 1953–54? If Stockman knew the answer, that was proof enough the messages he was receiving weren't planted.

Stockman happened to know the answer and told me to reply with, "Tell him the team leader was Dick Catledge and his wingman was Robert McCormick. There were twin brothers named Bill and Butch Patillo who also flew with them. Now, you tell that young lieutenant colonel down there to knock off this crap. We need him, but more importantly, he needs us. Tell him to get with the program."

Larson got with the program.

Bullshit

The North Vietnamese loved telling us how wonderfully we were being treated; how obliged we should be for their compassionate care; how much their people hated us, and that it was good we were inside the camp. Had we tried to escape the local citizens were inclined to dispense their justifiable vengeance upon us. All we needed to do was show "good attitude," follow camp regulations, and we would continue receiving humane and lenient treatment.

John Stavast, a rather tall, gangly guy, who stood six foot eight, with stooped shoulders, related to us how, after being served this line of manure during an interrogation session, he handled this nonsense.

He looked down at this little guard and said, "Bullshit."

The guard kept talking and John repeated, "Bullshit."

The third time John said bullshit, the guard looked up at him and asked, "What is this bullshit you keep saying?"

"That's what Americans say when they don't believe a damn thing they're hearing," he told him.

The guard became indignant, raised his fist, and said, "I forbid you to say that word."

"Bullshit."

Kevin's Hot Wax Job

Even months after our ejection, Kevin still suffered stabbing pains in his back whenever he tried to move. Many days all he could do was lie on his bed and moan. He couldn't even stand to bow when the guards came to the door. When the pain didn't abate, and after repeated requests for medical care, the guards finally administered their own version of a heating pad, using hot wax.

The procedure involved placing a chunk of wax in a flat, eleven-by-fourteen-inch pan, heating it over a flame until melted. When the wax cooled to a thin pliable gel, they flopped the wax onto Kevin's lower back around his sciatic. After the wax hardened, they peeled it off and placed rotten smelling seaweed on the affected area to draw out the pain. They did this six or seven times over a period of several months. It didn't work worth a tinker's damn, but it was their attempt to relive his pain.

As with any highly technical medical procedure there are risks, and hot wax had its downside. One time when the guards came to work their magic, Kevin was lying on his stomach, naked as a newborn babe. Again, they heated the wax and set it aside to congeal, but before it cooled and set properly, they flopped the wax onto his back. Unfortunately, only the outer edge had hardened. The middle, although appearing hard, was still a molten liquid.

The scalding liquid ran down along Kevin's sides to his groin, frying his privates. Kevin leaped off the bed and began hopping around and swearing as he pulled off the blistering wax. I had never seen him move so fast. Along with the wax came patches of hair and he yelled even louder. There Kevin was hopping around, holding himself, and screaming in pain, while Joe, Gary, and I laughed our ass off.

Popping Boils

Kevin had a terrible problem with painful boils and his sensitive skin made the boils intolerable. One persistent boil repeatedly broke out on his butt and wouldn't go away. Because he couldn't reach it, I helped him out. By applying pressure around the boil with my fingers, I popped it. After many months of practice I had perfected my technique and could really pop that sucker across the room. We used every opportunity to amuse ourselves.

Anyway, one night Kevin asked if I would pop the boil on his butt. As Gary and Joe sat watching, I decided to have some fun. Wrapping a tee shirt around my head and placing a pair of my shorts over my face like a surgical mask, I was ready to play the role of a doctor performing major surgery.

I set up my medical clinic in the middle of the room where the light was better to inspect the boil on his right cheek, as I recall. Kevin sat on one of the sawhorses used to support the boards for his bed with his fanny stuck up in the air. Gary and Joe razzed me about playing with Kevin's stuff, teasing us that we had the hots for one another. My effort to pop the boil was purely professional, of course.

Decked out in my surgical scrubs and with my thumbs poised, I positioned myself behind my patient on the other sawhorse, ready to squeeze the boil when the little judas door opened and an eye appeared. The door immediately slammed shut.

This was not good. Doing anything other than sitting on our beds was a violation of the camp rules and meant big trouble; we would surely be punished. But instead of bringing out the ropes, manacles, and foot locks, the door opened again, the guard stuck his hand through the opening, pointed to Kevin and me, and gave us the thumbs up. I heard him snicker as he closed the door. Who knows what he was thinking? But because we had seen guards walking together holding hands or with their arms around each other, we drew the conclusion that most of the guards were queer. We figured he must have thought he had caught us doing something naughty.

My First Christmas as a POW

In 1967 Kevin and I were taken to the Movie House to attend Christmas Mass. I'm not sure how they knew we were Catholic unless they garnered that fact from our dog tags which had been stripped from us after we were shot down.

Along with twelve other prisoners, we were ushered into the room and told to stand apart and keep our eyes straight ahead with our hands where the guards could see them. Photographers were everywhere, ready to record the event and show the world that we were allowed to attend a Christmas service. More propaganda bull.

The priest, who introduced himself as Father John Baptist, conducted Mass.

"God is going to get you for the evil you have brought upon Vietnam, and this is your punishment," he preached, haranguing us for bombing old folks, children, dam sites, and pagodas.

Yeah, mea culpa.

It was as though he were reading from a propaganda manual. I'm sure the interpreter added his own personal flavor to the priest's words, but I had to stand there with my aching back while he spoon-fed us Communist crap in the name of God.

I tuned out the priest's words and, for a brief moment, dared to think about Christmas at home. My father, keeping to the Mechenbier Christmas tradition, would line up all the kids on the stairs from the youngest to the oldest and make us wait until he took a photograph as we walked down the stairs to the living room to open our presents. I think he got a special delight in making us wait our turn, knowing we were excited to find what Santa had brought. Being the oldest, and last down the stairs, only made the wait all the more agonizing, but a pleasant memory when I looked back on it.

Christmas was the time we hauled out the Lionel train set and circled the tracks under the Christmas tree. Each year we added a newly acquired piece—a railroad crossing sign, a village house, or train car. We didn't have a big set, but using our imagination we pretended the train traveled to far away places carrying passengers and cargo.

Before putting the train away for the season, we linked the tracks through the house, winding them in and around the furniture, making the train's journey even more interesting.

Christmas was also a special time because that's when Mom made her wonderful glazed pastries with cherries on top. My mouth watered just thinking about them.

After "mass" we were treated to the Vietnamese tradition of Ban Chung Tet Cakes—a glob of meat in the middle of sticky, glutinous rice sprinkled with sugar, wrapped in banana leaves and steamed until the rice was cooked. We were each given a small cake, but with the bland meals we had become accustomed to, the sudden exposure to something so rich, even a small piece, made many men sick. To some, Tet Cakes were the only things worse than American fruitcake.

CHAPTER 8

Zoo
Pool Hall

THE ZOO, A FEW MILES FROM HOA LO, near the village of Cu Loc, covered several acres and was divided into a main area, an annex, and a section that housed the guards. A high wall separated each section. Buildings that housed the prisoners were made of heavy concrete with solid wooden doors and shuttered windows and tiled roofs. Many of the buildings, partitioned from other buildings by bamboo walls, had a colonnade that ran the length of the building. A long-abandoned, algae-filled swimming pool sat in the center of the compound.

Seeing chickens and pigs wandering around the grounds, the new arrivals called the compound the Zoo and dubbed the buildings the Barn, Chicken Coop, Stable, Pigsty, Garage, Library, and Pool Hall.

Kevin and I said goodbye to Joe and Gary in May 1968 with the hope we'd see them again and were moved to a place called the Zoo where we were housed in the Pool Hall 9. By comparison the Plantation had been a country club. Our eight-by-ten-foot cell, with concrete slabs for beds and cracked plastered walls, was half the size of the cell Kevin and I shared at Warehouse 13. There were no windows, and what little light there was came from two openings high in the wall near the ceiling.

The only redeeming factor about this move was Kevin was there with me.

The Fan Belt

Kevin and I got along relatively well despite living in tight quarters twenty-four hours a day. The only time we ever fought was to see who was going to light his cigarette first, and that was just silliness.

But there was one time—the only time—he got angry with me, and for good reason.

The summer heat made our cell unbearable day and night. I don't mean just sweaty hot; I mean lobster-boiling hot—a real sweatbox. With the outside temperature often peaking at 90 degrees, and with the extreme humidity and no ventilation, we sat in our cell baking like a Virginia ham.

The only relief we got was during a thunderstorm with a soaking rain. The guards, not having umbrellas and only pith helmets for protection, made themselves scarce, giving us an opportunity to relax. One of the more intriguing details about the rain was how the sound of the thunder rolled in the distance. You could hear it coming for miles. Those were times when Kevin and I would sit in silence, listening to the thunder and the rain pattering on the tile roof. The sound of the rain was not only soothing, it also brought slightly cooler air.

One night after cleaning our cell, we were lying on the concrete floor with our faces pressed to the space under the door trying to catch a breath of air when the judas door flopped down and a guard's face appeared.

"You have been communicating with broom. You will be severely punished."

It didn't take much to offend the guards. They were always riled up about something and took every opportunity to give us an "attitude adjustment," as they called it for our nefarious activities. We had, in fact, used our brooms earlier that day to send messages while outside sweeping the grounds. Seems we had been caught.

We were blindfolded and hauled out of the cell and led past cells numbered 8, 7, and 6. We took one step up into the bathing area, turned left, and made a half dozen steps around the corner to the courtyard. Another left turn, and another couple steps to the right put us in the Chicken Coop. I had been escorted along this path so many times I had the route memorized, even blindfolded.

Being taken to the Chicken Coop meant only one thing; communicating with a broom was frowned upon and a beating was forthcoming. They made Kevin and me kneel with our hands in the air while they read off a litany of our transgressions from their little blue book. If we didn't bow properly, they wrote it down. If we bowed holding our testicles, a definite no-no, in their book it went. Using stolen newspaper for toilet paper with Ho Chi Minh's picture on it really pissed them off, earning us a gold star beside the infraction. They kept very good records.

They accused us of making obscene finger gestures, which we did; using profanity and having a bad attitude, yep, we did that too; and not keeping our cells neat and clean and being slow emptying our buckets—guilty as charged. And don't forget Kevin peeing on the guard's foot.

I guess I should explain.

Kevin and I had only one, two-and-a-half-gallon latrine pot between us, and as you can imagine it soon overflowed. One night Kevin took the lid off the pot, knelt by the door, and angled it in such a way that his pee flowed under the door. As luck would have it, a guard happened to pass by just as Kevin let loose and he peed on the guard's foot. That little incident went into the blue book, probably in capital letters.

We knelt there for three or four hours while the guard ranted and fumed, telling us that we were going to be punished. It must have been three o'clock in the morning by the time he finally ran out of steam. Being Catholic, kneeling wasn't much of an issue, but holding my hands over my head for so long caused the blood to drain from them and turned them to lead.

We never admitted to any of their charges, of course, but convinced otherwise, the guard led Kevin and me out of the Chicken Coop and took us to the administration building. There I was made to strip, and ordered to lie face down, spread-eagle on the floor while four henchmen held my feet and arms.

Psycho, the sadist, who was never allowed to be alone with an American for fear he would "lose control" in his frenzy to inflict pain, took an eight-foot-long piece of rubber fan belt and began whipping it across my back and butt. As he thrashed me, the Vietnamese who weren't holding me, stood and watched, I guess to make sure I didn't fight back.

I had lost a lot of weight during my imprisonment, and when the guard whipped the broad part of the fan belt across my backside, the tip of it hit the floor, making a tremendous snapping sound against the concrete but inflicted only a mild discomfort. When I realized that the belt wasn't long enough to do much damage, I feigned pain and began yelling and screaming as though they were killing me. They shoved a dirty rag in my mouth to shut me up and kept on whipping. Only after Psycho became exhausted did he stop.

When it came Kevin's turn for his attitude adjustment, I whispered to him not to sweat it because the belt was too long and wouldn't hurt. "Just yell like hell," I told him. Although we weighed about the same at the time we were shot down, I hadn't taken into consideration that Kevin still had his baby fat and was larger than me. With each lash, the belt cut into him leaving deep, red welts.

Kevin couldn't sit down for two weeks. That was the only time he became distant and didn't speak to me. I knew the belt had hurt him and I was truly sorry and apologized.

"You just got even with me, that's all," he said.

"What do you mean?"

"Remember what happened to you that first day after we were captured?"

"Yeah, nothing. I heard you getting knocked around down the hall, though."

"Well, they did more than knock me around. They had me in that Knobby Torture Room beating the crap out of me when it dawned on me it was you they wanted. I told them I was just a backseater, a nobody. They needed to talk to the aircraft commander, the guy down the hall."

Perhaps because Kevin was bigger and looked older, they assumed he was the pilot flying the airplane. I guess you could say he got even with me, or maybe I got even with him. Either way, we were even.

The Solid Stool

Having become experienced distributors of communications while dishwashers at the Plantation, Kevin and I were appointed to continue the job at the Zoo, guaranteeing that messages reached their destination. In an effort to catch up on the latest news and gossip, we employed our sophisticated communication system and usually "talked" during siesta between noon and 1400 hours when fewer guards were around.

The men who had fellows living in a cell next to them could tap on the wall to communicate, but to send and receive messages from another building we had to use hand signals, and the only way to do that was through the tiny opening high in the wall near the ceiling. That meant standing on your cellmate's shoulders.

As mentioned before, we always began communicating by tapping the old "shave and a haircut." But if the message was really urgent and needed to be transmitted immediately, the sender would tap rapidly to get everyone's attention.

Having mastered the hand code, I could anticipate what another POW was going to say even before he signaled it. But I wasn't prepared for the message I got one Sunday morning when the Comm Net went up on high alert with a rapid, almost frantic, tapping. Something important was up.

I climbed onto Kevin's shoulders, looked out the tiny four-by-four-inch vent hole near the ceiling and watched carefully as the message was flashed by hand signal from the Stable across the way. My job was to forward the incoming message from my cell in the Pool Hall to Don Burns and Ted Kopfman living in the next cell, who would then tap the message to Larry Guarino, the SRO, who lived next door to them. The message having come from the Annex had to travel a great distance as it came over the wall to the Garage to the Barn to where it was then transmitted to the Library on to the Pigsty over to the Stable and then to us.

The message: "Jack Fellowes had a solid stool."

Once received, Guarino asked, "What did you do with it?"

The tapping began again, but in reverse from Kopfman and Burns to Kevin and then back to me. I would then whistle to alert the men in the Stable that another message was coming to them. I also watched for guards who might be wandering around close by. When the message was received in the Stable, it traveled back across to the Pigsty to the Library over to the Barn and then to the Garage where it was sent back over the wall to the Annex.

The communication took nearly an hour to reach the original sender while 60 to 70 guys hung out their you-know-whats to forward the message.

The answer: "We looked at it."

A solid stool was big news, considering that most everyone suffered from dysentery. Then the question was raised: Was this the first solid stool anyone ever had while here? To amuse ourselves, we took a poll. The survey concluded that it was, in fact, the first solid stool anyone had had in sixteen months. That was even bigger news.

The situation became even more laughable when we started receiving packages from home. Air Force personnel, in all their wisdom, sent us pills for constipation. Here we were suffering with internal parasites, doing the green apple two-step, and they're sending us laxatives. I must offer a word of thanks to our military intelligence.

First Package From Home

In March 1969 I was ushered from my cell and taken to the dreaded Chicken Coop. Suspecting the worst, I entered the room with trepidation, but was relieved when told I had received a package from home. For some reason I was one of the last prisoners to get a package. As I entered the Coop, I spied a washcloth, a pair of undershorts, two bourbon balls, a small box filled with what looked like Red Hots candy sitting on the table, but more importantly, there was a photo of Jerri.

That was it. Packages were allowed to be up to 6.6 pounds. What I saw lying on that table didn't amount to diddlysquat. The Vietnamese had pilfered most of the items because there was no way that package weighed six pounds with just these trifling offerings.

Then came the quid pro quo. In order to have the package, they wanted me to sign a paper stating that I had received it because I was grateful to the Democratic Republic of North Vietnam for their merciful treatment.

I wasn't about to sign any such agreement.

We dickered. They wanted me to sign the paper; I wanted the package. After hours of wrangling, I had to pee, but with no way to relieve myself, I sat there unwilling to give them the satisfaction of soiling myself. Finally, with my bladder ballooning, I said I would agree to sign, but only for the items received.

Nope. That wasn't good enough. We continued haggling

As each hour passed, my bladder problem became more and more difficult to ignore. We went around and around, until finally they were satisfied. I signed a paper that ended up being nothing more than a receipt for my package, but not an acknowledgment of their charitable treatment.

By the time I got my package and stood to leave, my bladder felt as though it would burst and I could barely walk. All I could do was hobble, doubled over, back to my cell cupping my few treasures in my hands.

Looking through their vent holes, other POWs saw me hunched over, shuffling toward my cell with a guard following close behind with his bayonet pointed at my back.

Rapid tapping like that of demented woodpeckers on crack cocaine flew through the camp. "Mechenbier got the crap beat out of him for his package."

Once back in my cell, I went straight to the corner and knelt by the bucket, but I couldn't release. It seemed to take forever for the flow to start, but when it did it was like the flood gates of heaven had opened.

Afterward, I turned my attention to the items Jerri had sent. I shared the

Red Hots with Kevin, I tasted one and savored its hot, tangy flavor. I tried another, but it tasted stale and flat. Kevin agreed that some of the Red Hots tasted funny. After eating all the candy, we discovered that Jerri had mixed in red One-a-Day vitamins. We were on a vitamin high for days.

The heavenly aroma of liquor floated through our cell when I opened the bourbon balls. Even after weeks in transport, and wrapped in foil, they still held their rich flavor. With each bite we basked in the effects of the aphrodisiac. Oh, the joy! For a brief moment in our dank, dreary cell it was Happy Hour, Hanoi style.

A Very Special Cigarette

We were often subjected to unannounced searches of our cells. Camp rules dictated that we keep our space neat and clean. Failure to do so gave the guards reason to point out the error of our ways with a swift kick in the rear or a rifle butt across our back. While we were emptying our buckets, picking up our food, or taking a bath, guards would tear through the cell, looking for contraband, and making a holy mess in the process.

What the guards didn't take from our packages from home, they'd steal from our cells during inspections. They stole anything that was a source of enjoyment including our playing cards we had painstakingly made from six or seven layers of newspaper stuck together with a plain piece of paper on one side and a crudely printed number on the other. They also took the dice we had made from little pieces of bamboo. These possessions weren't much, but we cherished them, and even though we were crafty in hiding our valuables, the guards were quick to find and confiscate them.

Pocks, one of the meaner guards, loved stealing our cigarettes. For most men, a cigarette was the only luxury they had in this pit, if you could call smoking cigarettes containing rat turds a luxury. To us, taking our cigarettes was a covert form of punishment, and we took the abuse for just so long before we got fed up with Pocks's total disregard for our only pleasure and rebelled. We devised a plan to thwart him.

The wrapper on Vietnamese cigarettes had a tea-stained color, making it easy to put our plan into action. We took a cigarette from our stash, put it in the honey bucket until it soaked up the flavor, and then set it out to dry. Later, we put the cigarette back into the pack with the others, and then sat back and waited for Pocks to help himself. We never had a problem with him taking our cigarettes again.

SIUTGA.

Loud Speakers

Upon arriving at the Zoo, I noticed the Vietnamese had loud speakers placed outside the camp that blared from dawn to dusk. Each day began with gong, gong, gong telling the umpteen million Vietnamese living in the city that it was time to get up. Along with news and sports, Ho Chi Minh style, they played the same piece of music all day, which sounded like a Rachmaninoff concerto. Then in the evening the gong, gong, gong sounded again letting everyone know it was time to go to bed.

There were also loud speakers strategically placed throughout camp, and every night we were bombarded with a half-hour broadcast of the *Voice of Vietnam*, first in Vietnamese, and then we'd hear another half-hour version of their twisted propaganda and distorted news in English. That was followed by a dose of Hanoi Hannah giving her rendition of the war.

The program was nothing more than an attempt to brainwash us into believing their propaganda crap and to feed us garbage about North Vietnam's struggle for independence and reinforce their position that we were war criminals. They attempted to indoctrinate us by playing endless hours of tape recordings recounting the battle of Dien Bien Phu. That was followed by a quick course on Vietnam's history. If that wasn't enough, we were subjected to the rants of antiwar activists such as Jane Fonda and Ramsay Clark trying to demoralize us by claiming we were black-hearted criminals.

The Vietnamese also loved yanking our chain by playing the same three songs over and over. We heard hours of "Going Home" an old World War II song and Nancy Sinatra's "These Boots are Made for Walking." The guards also made repeated reference to our wives having affairs, getting a divorce and remarrying, and I'm sure they got a big kick out of the verse from the song "Cecilia" made famous by Simon and Garfunkel that said, "When I came back to bed someone had taken my place." That phrase was a real punch in the gut.

We couldn't avoid hearing the speakers throughout the camp and one day a guard asked me how I enjoyed the music. I told him I couldn't hear the music because I didn't have a speaker in my cell—end of discussion.

SIUTGA.

The next day, my cell door opened and in walked two guards carrying a squawk box, which they promptly installed in the corner of the cell. Now, we could hear their broadcast up close and personal and at the same ear-piercing volume as the large speakers outside in the yard.

SIUYOA.

The Slogan Contest

When the Paris Peace talks began in May 1968 the Vietnamese were prompt to broadcast a blow-by-blow account of the proceedings. While they kept telling us that victory would soon be theirs, we started a betting pool that turned into a slogan contest as to when we would be released.

The talks stalled when the delegation couldn't decide how many participants there should be, whose flags would be displayed, or if square or round tables would be used. This crap went on for six months as they argued over every detail.

In 1968 the winning slogan was, "Golden Gate in '68."

The wrangling continued as the delegation tried to work out an agreement as to whether there would be flowers on the tables, what size flags to use, and who would be introduced first.

In 1969 the winning slogan was, "Everything's fine in '69."

The talks dragged on with no resolution in sight. Now they couldn't decide on the official language to be used—French, Vietnamese, or English?

Finally, in 1970, the winning slogan was, "Ah, screw it."

The U.S. ambassador-at-large, Averell Harriman, left the Paris Peace Accords empty-handed after talks with governments from the Democratic Republic of Vietnam (North Vietnam), the Republic of Vietnam (South Vietnam), and the United States, as well as the Provisional Revolutionary Government (PRG) stalled and no agreement was reached. The war continued.

Middle Finger Salute

After emptying my honey bucket one day, I was walking back to my cell when I saw another American heading in my direction. The guards had obviously screwed up and got their signals crossed and had allowed two of us out of our cells at the same time—something they never did.

Saluting or extending my hand to my fellow American would have resulted in a kick in the back from the guard, but being the polite guy that I am, I discretely flipped him the bird.

The guards, seeing me give a fellow prisoner the finger, wanted to know what it meant. I told them that extending the middle finger with a raised hand was a salutation American fighter pilots used when greeting each other in friendship. We had already taught them the "F" word and told them that if we really liked someone we'd say, "Screw you," with a raised middle finger. They must have thought that was pretty cool because the next time a guard came to the door, and wanting to be like the Americans, he flipped us the bird when we bowed.

Later, we saw guards give each other the finger along with the salutation, "Screw you," as they walked their beat in the courtyard. It was hysterical watching them try to imitate the American reprobates who had fallen from the sky. We had a pretty good run with that for about three years.

SIUTGA.

Word of the new American salute quickly spread to other guards at other camps, until three young ladies from the Students for a Democratic Society, part of an antiwar delegation, visited the Plantation. The guards, wanting to be cool, flipped them off. The women were humiliated, as you might imagine, and explained what the gesture actually meant.

Learning the real meaning behind the gesture, the camp officer nailed Tom Parrott as the instigator. Apparently, they went back through information documented in their blue book and found that Tom had once made the gesture. Therefore, he was to blame and was put in isolation.

SIUYOA.

The Hair Pulling Incident

Our cell was always stifling hot during summer and the simple gesture of moving our arms and legs took every ounce of energy we could muster. One night, Kevin and I were lying under our mosquito nets stark naked, trying to find relief from the heat. We had our arms and legs spread out with our shorts over our eyes to shut out the constant glare of the light bulb when the judas door opened. Seeing us lying there without any clothes, the guard became enraged and began yelling and screaming.

He then stuck his hand through the judas door and motioned for me to come closer. I started to get up and go to the door, but first I had to climb out from under my mosquito net.

Getting in and out of bed from under a mosquito net takes skill, agility, and cunning, and if done correctly, it is a real art form. It took me months of practice to master the technique. First, you had to lift the net just high enough to slide your legs off the bed before bugs got in. Next, you needed to throw the net up and over your head, and then quickly drop it behind you. To get into bed, the process was reversed except for the extra step of waving your arms in the air to shoo away the countless mosquitoes that might be hovering close by. You had to act fast as you slipped under the netting before tucking it in around you. Like I said, it took practice.

After successfully getting out from underneath the netting, I made my way to the door. As I got within reach of the guard and bowed, he grabbed my hair and started pulling, still yelling in Vietnamese. I didn't have to know the language to understand that I had done something that pissed him off.

Hearing all the noise, Kevin woke up and looked around and asked, "What does he want?"

"Come on over here. The guard wants to pull your hair."

Still half asleep and bleary eyed, Kevin got out of bed and started toward me. Half way to the door he stopped, looked at me, and replied, "No, I don't think I want to do that," and then crawled back into bed.

The guard was having a fit and Kevin just blew him off. That was so like Kevin. I laughed until I had tears in my eyes.

The guard seemed pleased that he had made me cry.

CHAPTER 9

Zoo Garage

THE GARAGE, A LONG BUILDING WITH SIX CELLS, sat at the far end of the main compound, separated from the Annex by a high, stone wall.

That summer Kevin and I were moved from the Pool Hall and taken across the compound to the Garage. We were put into cell no. 4 where we met Ron Bliss, a short, balding gentleman who had a habit of looking at you as if to say, "you gotta be kidding," and Galand Kramer, who we called "Lurch" because he reminded us of the TV character on the *Addams Family.*

Our cell was slightly bigger than the one in the Pool Hall, but our beds, like those in the Plantation, were wooden slats laid across two sawhorses rather than concrete slabs. The cell had a big double door at the end of the room, and despite the larger, louvered windows, there was still very little ventilation.

The Great Escape—Mad Dog Summer

The summer of 1969 became known as Mad Dog Summer after Edwin Atterberry and John Dramesi attempted an early morning escape from the Annex only to be recaptured a couple hours later. That's when the crap hit the fan. The Vietnamese went stark raving mad, unleashing their vengeance, and turning our lives into hell on earth. With the threat of death hanging over us every minute, we lived in constant fear.

I can understand Atterberry's and Dramesi's desire for freedom and their attempt to escape, adhering to the Code of Conduct, but with no assistance on the outside, their escape plan was futile.

The escape ended tragically for Atterberry who was tortured to death. Dramesi was placed in irons for six months and his eight cellmates were beaten unmercifully.

While torturing the men in the Annex, the guards ascertained that others had been communicating with Atterberry and Dramesi and had formed an escape committee. That sent waves of terror throughout the rest of the camp. We lived in fear that the scourge would reach us, which it did.

Before the escape, if a camp officer wanted to question or correct our behavior, we were taken to one of the interrogation or torture rooms and punished. Now, they brought the beatings to our cells. It was as if someone had turned them loose and gave them free rein to do whatever, whenever, wherever they wanted.

First they cut off our only chance of fresh air by sealing the air vents, making living conditions even more miserable. Then the guards began controlling every facet of our lives. They made us line our cups on the end of our beds with the handles pointing in one direction. If the handle wasn't turned exactly as they ordered, we'd get the crap beat out of us. If we didn't stand quickly enough or bow properly, we got a rifle butt across the back. If our blankets weren't folded exactly into thirds, we got jabbed in the gut. If our cell wasn't as clean as they expected, we got punched in the face. This relentless sadism continued for three months.

Although we always had the feeling that the guards were trying to provoke a fight, there was never a situation where one of us became angry enough to strike a Vietnamese guard. That would have been a death sentence.

Rats, Geckos, Spiders, and Snakes

Not only did we have the same variety of insects—flies, cockroaches, fleas, and lice—as we had at the Plantation, here at the Zoo we also had rats, geckos, spiders, and snakes to contend with.

Surrounded by marshy lowlands and rice fields, during the rainy season, the Zoo became a warm, comfortable haven for the slithering and gnawing creatures seeking shelter from the rising floodwaters. It was a virtual breeding ground for every conceivable creepy-crawly that called the Zoo home.

Rats in Vietnam were as common as warts on a frog and caused little concern to the guards. Rats were just part of their everyday lifestyle, like a pet dog is in ours.

Vietnamese rats were big and weren't afraid of humans. Accustomed to free reign, they walked around as if at a Sunday picnic, taking little notice of us. It wasn't unusual to hear yelling in the middle of the night as a rat crawled into bed with one of the guys to snuggle up to a warm body.

As I was being led from my cell one evening, I encountered three rats, each about the size of a Schnauzer, scurrying down the rain-filled trench in front of the cell door. I waved my hands above my head, stomped my feet and made a hacking sound that would have scared any normal God-fearing animal, but these rats just looked up at me as if to say, "You got a problem, Mister?" and kept on going.

A wide variety of small critters also showed up regularly to nibble on anything they could find in our cell, but our favorite guests were the geckos that crawled around on the ceiling. They didn't bother us; in fact, they were quite helpful in controlling the mosquito and fly population. They became a source of amusement, as we'd sit on our beds watching them lay in wait for a tasty morsel. Since we had no choice but to share our cell with them, we made the best of it, besides they didn't want our food.

While living in Albuquerque, when my brother Jim and I weren't saving the world from bad guys with our cap guns, we'd often play in the field near our home catching toads and lizards. We'd fashioned little harnesses out of pieces of string and wrapped them around the legs of blue-tail lizards and kept them in cages as pets.

Using my acquired knowledge of lizard harnessing, Kevin and I passed the time fashioning miniature halters for the geckos out of string we pulled from our blankets. We became quite good at designing harnesses in such a way that they fit around their tiny bodies and behind their front legs. Who would have ever guessed that something I did as a kid would one day come in handy in a Vietnam prison cell? For that matter, who would believe that an Air Force Academy graduate would spend so much time playing with geckos?

We had an abundance of spiders, which Kevin hated. Vietnamese spiders were unlike ordinary America household spiders. These were big, black, and hairy, about the size of a human hand, and they came out only at night.

As a rule, when the air raid sirens sounded, we stayed in our beds until the danger past, but one night along with the sirens we heard a loud explosion as a bomb landed close to camp. Hearing the bomb, I threw off my mosquito net, sprang out of bed, rolled under it, and wrapped my arms around my head for protection from any falling debris. Kevin, still half asleep, also rolled out of his bunk, hit the floor, and crawled under his bed.

The only time the lights were ever turned off was during air raids, so the room was pitch dark. As Kevin lay there, he remembered the spiders. In a panic to avoid the creepies that may have been lurking under his bed, he jumped up, causing the boards to fly off the sawhorses, and crash to the floor. The guards, apparently thinking we were trying to escape during the bombing raid, came running, guns drawn, but all they found was Kevin laying on the floor cocooned in his mosquito net.

The rainy season spurred boatloads of snakes from their nesting holes, and never knowing which snakes were friend or foe—nonpoisonous or "stay away from me"—we brave knights of the air showed a guarded respect for them. We saw them routinely, and to provide the greatest margin of safety, we encouraged them to slither elsewhere. Our abhorrence of snakes outweighed our desire for fresh air, and even though it was often stifling hot in our cells, we would stick clothing in the slit under the door to keep the reptiles out.

Ron Bliss and the Orange Booze

Kevin, Ron, and I, being 1964 graduates of the Air Force Academy, held the same date of rank. Galand Kramer, however, was an ROTC graduate from Oklahoma State University. By tradition, ROTC cadets always graduated a day after the Academy, giving the three of us a one day advantage. Because Kevin, Ron, and I had the same date of rank, we needed to decide who was the senior ranking officer, and according to procedure, went alphabetically, making Ron our SRO.

To celebrate Vietnam's National Day on September 2, the guards gave all of us a couple ounces of very sweet orange liqueur. Agreeing it was better for one guy to have a good buzz than for all of us to get only a little tingle, we made a game of it and held a lottery. We drew straws and rigged the contest to make sure Ron won.

Having gone without alcohol for years, and not being in the best physical shape, our bodies could not tolerate even the smallest amount of alcohol. When Ron swigged the several ounces of liqueur on an empty stomach, the booze hit him hard. For the next several hours all he could do was sit on his bunk, staring into space, goofy eyed—drunk as a skunk.

Bathing in the Cesspool

It didn't happen often, but on a rare occasion a guard did something to demonstrate the warm-hearted treatment we were supposed to be receiving. Being allowed to bathe wasn't considered a special favor, so Kevin and I didn't balk when Pocks allowed us to bathe in the pond behind the Pigsty. The turbid water was covered with algae, but it was cool.

We splashed and frolicked in the water several minutes before Pocks motioned for us to climb out and returned us to our cell. Once back in our room the tapping began: "Who were the two idiots out there in the . . . pond where we dump our buckets?" The guys must have had a good laugh peering through their peepholes.

The next day Kevin and I began itching and broke out in red spots all over our bodies. A few days later our skin turned yellow and we began to feel like crap, with flu-like symptoms. We suspected we had contracted Hepatitis A, B, and C and probably the rest of the alphabet. That's when we knew that Pocks had stuck it to us for his specially crap-dipped cigarette.

SIUYOA, boys.

Rx Iodine

Living in a constant steam bath during the summer was brutal, and it wasn't uncommon, without the means to bathe frequently and without talcum powder, for many men to get a bad case of jock rash.

The Vietnamese remedy was Iodine. Boxi, the Vietnamese word for nurse, would come in, line us up, order us to drop our shorts, and tell us hold up our genitals. Using a cotton ball soaked in Iodine, he would then swab up one side of our crotch and down the other. It burned like hell but eventually soothed the constant itching and irritation.

The Vietnamese used Iodine as a cure-all for everything, and sometimes the cure was worse than the problem. Bob Biss, an exercise nut, had just finished doing several sets of jumping jacks and had worked up a good sweat; he also had a raw spot on the end of his penis. While Boxi was taking care of Bob's cellmate's jock rash, Bob showed the guard the irritated spot. Being an accommodating fellow, Boxi swiped the Iodine across Bob's privates. We could hear Bob's agonizing screams all the way across the camp.

A Vietnamese Christmas Dinner

For Christmas 1969 the camp officer told us to expect something special for Christmas dinner. Knowing that everything the Vietnamese did had a political motive and could be used for propaganda, we took nothing they offered at face value. There was always a catch.

Looking through the peephole in the door, we saw a turkey, potato soup, bread, lychee nuts, and chunks of meat chopped into squares laid out on a long table outside our cell. Photographers stood with their cameras ready to show the world our delight and appreciation for the wonderful treatment we were receiving.

Cameras! That was the first red flag that something was up.

The guard told us this was a special day and we didn't have to bow when we went to get our food.

Another red flag!

The four of us weren't going to allow them the satisfaction of using us as propaganda puppets. To derail their scheme, when the door opened, we walked out like zombies, sucking in our cheeks, dragging our feet, with our arms and bodies bent and contorted. I took the plot one step further and stuck my tongue out the side of my mouth and drooled. We hobbled to the table, picked up our food and pretended to have difficulty eating.

The commander was livid and screamed, "Act right! Act right!"

We were promptly forced back to our cell as the commander yelled, "Bow! Bow!"

Once back inside, we broke into fits of laughter. We had outwitted them and had successfully ruined their propaganda film. It was a little victory, but it was ours.

SIUTGA.

The Christmas Skit

No matter how teary-eyed some might have become during Christmas season there was always someone who found a way to turn an otherwise gloomy moment into something funny.

Quincy Collins, who had been a training officer at the Air Force Academy, and who had a wonderful singing voice, was asked to produce a Christmas skit to be taped and played for *our* entertainment, knowing full well that it would be played over the *Voice of Vietnam*.

Quincy questioned making the tape but after obtaining approval from the SRO, went ahead with the project. Although he knew the Christmas skit would be used for propaganda, he figured producing the program and burying names within the song would be a good way to secretly get word out about the men being held prisoner. For example, he revamped "Rudolph the Red Nosed Reindeer" to "Benny the Pride of Elba," hoping that his family in Elba, Alabama, would understand that Ben Ringsdorf was alive.

Using the melodies of familiar Christmas carols, he turned them into irreverent songs such as the Sandman dropping his load, and Santa Claus coming down the chimney ass first making brown streaks in his shorts.

He got away with the scheme for several months until the tape was broadcast to an antiwar group back in the states who understood the subtlety of the words and exposed the double entendre to the Vietnamese. The choir was summarily disbanded and we no longer were allowed to sing during our church service.

SIUYOA.

Night Out at the Museum

In the early spring of 1970, a guard gestured for the four of us to put on our long sleeve Mess Dress pajamas. We were blindfolded, put into the back of a truck, and driven downtown to the Army War Museum where we were taken into a large room and cautioned not to speak to anyone—just sit, listen, and learn.

Surprisingly, there were no cameras anywhere.

The lights in the room were lowered and for the next hour we were shown a documentary film on Vietnam's struggle for independence. The featured event was the recreation of the Dien Bien Phu battle of 1954, illustrated in great detail through a diorama that demonstrated the battle scenes. Lights projecting up through the bottom of the display illuminated thirty square miles of trails that the two forces followed through the mountains. Flags moved up and down, showing the Vietnamese colors, as they advanced up the hill into the fort to defeat the French.

After the demonstration, we were once again herded back to camp and locked in our cells. That was it. The trip to the museum served no purpose except perhaps to teach us how the Democratic Republic of Vietnam was formed in case the U.S. educational system had failed to enlighten us.

I had expected them to question us as to what we had seen and if we had come to a greater appreciation of their country, but they never made an effort to test us. It was just a night out on the town sitting in a much cooler place than our cell. Whoopee!

Our education didn't end there. We were often taken into a darkened auditorium on the compound to watch blurry antiwar films, newsreels, and war movies. It was laughable watching movies that showed the Vietnamese shooting at American airplanes and seeing the same film sequence over and over to make us believe they had shot down a hundred airplanes.

It was a low budget production. No cartoons.

CHAPTER 10

Camp Faith

DAN HOI, ALSO CALLED CAMP FAITH, about ten miles west of Hanoi on the main route between Hanoi and Son Tay, sat nestled in the countryside, surrounded by farmland.

The large compound was divided into six separate sections each having its own guard tower, courtyard, and a separate latrine and bathing area. Each building, made of hewn stone and stucco, had five cells large enough to hold eight to twenty prisoners. The camp was surrounded by a sixteen-foot-high wall covered with barbed wire.

In July 1970 the Vietnamese played musical chairs, but with prison camps. I was transferred to Camp Faith where I shared a cell with Bob Wideman from Bay Village, Ohio, who happened to be the world's loudest snorer. He kept me awake many nights with his rendition of a sputtering chain saw. The guards suspected him of communicating by snoring in code.

Also included in the mix were Dave Luna, a bit of a cutup and a Wild Weasel from Washington, D.C.; Wayne Smith, a 1965 Air Force Academy graduate; and Ben Ringsdorf, who holds the world's record for the stupidest puns. Then there was Ron Bliss, who you might remember from his booze drinking days. The other two gentlemen were Joe Milligan and Gary Anderson, with whom I had previously shared a cell at the Plantation in 1967.

What a group that was. For some reason, they didn't move Kevin with us, and I had no idea where he was.

Our twenty-by-twenty-foot cell that appeared to have been recently whitewashed had two barred windows on the back wall that opened to allow sunshine in during the day and a cool breeze at night. Our beds were still boards laid across sawhorses, but here the wood was teak instead of bamboo.

The environment at Camp Faith was quite different from other prison compounds where I had been held. The guards seemed more congenial and went about their business without bothering us. Being in the countryside where we could breathe fresh air and hear birds singing, we experienced a more relaxed atmosphere without much interference from guards. But more importantly we were spared the tortuous hours of interrogations and constant fear of being beaten. It was a treat to be allowed outside longer than fifteen minutes a day to mingle and talk face to face with other Americans without having to resort to subversive means of communicating.

When the guards wanted to be nice, they allowed us to take our time bathing in warm rainwater that ran off the tin roof of our building and collected in a cistern. No more icy water. We received better meals, not great mind you, but better than the slop we got at the Hanoi Hilton, Plantation, and Zoo. Here, along with the regular fare of pumpkin soup and stale bread, we occasionally got fruit and meat.

Needles and Novocain

Medical care was better, and men finally got treatment for long-neglected wounds and injuries—minimal as it was—giving hope that the more serious injuries would be addressed. Although some complaints were minor, others were acute, needing comprehensive medical attention. Unfortunately, the request to see a doctor took days, sometimes weeks or months, before Boxi paid a visit.

Jeff Ellis, being held in a building across from ours, had a wound on his leg that never seemed to heal and kept festering. After three and a half years of complaining, they finally sent someone to take care of him.

I watched, peeping through the window of my cell, as Needles, the "doctor," looking every bit the medical professional in his doctor's hat, facemask, and surgical operating scrubs, appeared at Jeff's cell door. Novocain, his assistant, hovered close behind holding a tray of instruments.

"Now, how it is that anyone needs help today?" I heard Needles ask, looking around the cell. Not waiting for an answer, he motioned for Jeff to come closer and in a most serious manner, he looked at Jeff's leg and said, "Ah, yes. Must clean wound." He knew exactly who he was there to see, and why.

Needles uncovered a jar of cotton balls sitting on the tray, reached in with his fingers, picked one up, dipped it in alcohol, and then cleaned his hands with it. Sterilized and ready to perform his medical miracle, he raised his hands in the air, taking on a preoperation pose.

Realizing that an instrument was missing from the tray, he sent Novocain to fetch the item. Novocain scurried from the cell, leaving Needles standing with his hands in the air. Everyone waited—five minutes, then ten. Needles must have gotten tired because I saw him lower his hands, and stuffed them under his armpits.

Needing a smoke, he pulled down his mask, reached into his pocket—who knows what might have been growing in there—and brought out a cigarette, lit it, and sat outside on the stoop smoking. As a dog walked by, Needles gave him a friendly rub behind its ears.

When Novocain finally returned with the missing instrument, Needles slid his facemask over his mouth, put his hands back in the air, and resumed his sterile, preop position. Ready to go to work at last, he clamped a cotton ball between the teeth of the forceps and stuck the ball into the alcohol jar, and then swabbed the wound, but when he expanded the forceps to remove the cotton, it wouldn't come off. He shook the forceps but the cotton still wouldn't budge. Looking around and seeing the honey bucket in the corner, he tapped the forceps on the rim of the bucket, picked up another cotton ball with the same forceps, dunked it into the alcohol, and then started back to work on Jeff's leg.

Naming the Guards

To pass time, we often sat around comparing notes about guards we had encountered in various prison camps and discovered we had given many of them names based on their resemblance to an animal, cartoon character, or because they possessed a specific peculiarity. Some guards had acquired several names according to who had applied the moniker.

All guards dressed in the same drab tan uniforms, and like the prisoners, wore rubber sandals. In my early days of imprisonment, except for the red star on their pith helmets, none wore markings on their uniforms to signify rank. But later I noticed some interrogators wore bars on their shirt collar and white stars on their helmets.

Rabbit, who qualified as a circus sideshow reject, was the most sadistic interrogator. He was the camp commander at Hoa Lo, and a vain SOB, a real egomaniac. He spoke perfect English, at least in his own mind, and thought nothing of reaching across the interrogation table and slapping a prisoner.

Soft Soap Fairy, aka, Ensign Parker, and Frenchy, who I mentioned earlier, was most likely an officer. He had a slight build and was always well groomed. He was the suave interrogator, fluent in French, and had a good command of English. Unlike the other officers, he wore a brown tunic and polished brown shoes, not sandals. Interestingly, he never hung around while we were being tortured. During one of his interrogations, he told me that if I didn't cooperate there were those who had ways to persuade me, and then he left the room. The persuasion followed.

One older, senior interrogator, tagged Bug because of his large insect-like eyes, always had a cigarette dangling from his mouth. Pocks, the cleat face, thought nothing of lighting your cigarette one minute, and then without hesitation burning you with the punk stick as soon as your back was turned.

Then there was Wug, an acronym for the World's Ugliest Goon, and Sug, Son of the Ugliest Goon. Both were bug ugly. Slick, a medic and cohort of Needles and Novocain, wore his dark hair combed back. Rigor Mortis, also known as Psycho, was a sadistic goon who took delight in hurting people. Once we watched him twist a dog's leg just to make the poor thing squeal in pain.

Boo Boo was the personification of Boo Boo in the Yogi Bear cartoon and was always in a good mood, smiling and never mean.

Stag—Smarter Than the Average Guard—conducted his interrogations on the stoop of our jail cell. Stag had a dialog with one of my roommates at the Garage in the Zoo that went something like this: "Look at the sandals I'm wearing. They are not as good as the boots you were wearing when you got shot down; and the uniform I'm wearing, the material is not as good as the material in the flying suit you were wearing. You know how I got these things? Our Russian and Chinese brothers gave them to us. That's why we are fighting—to get better stuff."

He was the only guard to whom we could complain. The bread they gave us showed signs where rats had already eaten their share. We told Stag that the bread needed to be sealed tightly in the bin to prevent rats from getting into it, otherwise we could get sick from disease. Taking our suggestion, he actually had the bread crib lined with tin. He even took one of the POWs to the kitchen and showed him how they had indeed fixed the problem.

After Ho Chi Minh died (September 3, 1969) Stag gave us a chess set and Russian literature—Tolstoy, Chekhov, and Dostoyevsky—that he let us keep a day or two. He also gave us a deck of playing cards.

In a conversation with Kevin, Stag made the comment that he would soon be leaving because he was needed at the front. We never saw him again and assumed he was fired for being too friendly with prisoners.

We also had names for a couple of the Vietnamese women who worked in the kitchen. While at the Plantation, we observed them carrying water and cooking utensils across the compound. One we called Pocahontas because she looked like an American Indian. Some of the guys fantasized having her as a bunkmate. The other woman we called PU—Princess Ugly.

What Fidel Thought of the Guards

The Cuban Program was an extensive operation instituted in 1967 to extract information from Americans that the Vietnamese seemed incapable of obtaining. If the American prisoners considered the torture they received at the hands of the Vietnamese harsh, they soon learned that it was child's play compared to the torment they suffered under Fidel's sadistic tactics.

Americans weren't the only ones who thought little of the Vietnamese. Even Fidel, a name given to an interrogator who came to Vietnam to institute his Cuban Program, had no respect for them.

While visiting the Zoo a year earlier, Fidel made his contempt known. Kevin and I had been given the task of driving the blunt end of a U-shaped bolt into a piece of wood with a brick. We didn't care that the job was ridiculous; it got us out of our cell for a few hours. As we were pounding away at the bolt, Fidel walked by and asked what we were doing. We told him.

"These people are bug-—— stupid aren't they," he said.

Ron Bliss and Ben Ringsdorf had a similar experience. They were told to move cement blocks being used to build a new bathhouse behind the Library at the Zoo. Each time they picked up a block it shattered and fell apart. Noticing the men fumbling with the blocks, Fidel said, "These people make crappy cement," and walked away.

Kevin and I also learned first hand how shoddy their construction materials were. The Vietnamese used concrete cylinders in the building of their air raid shelters and wanted us to move the cylinders from the gatehouse around the pool behind the Stable. We didn't want to help, but knew if we didn't they'd clobber us. So, the question then became should we roll the cylinders and violate the Code or refuse and take the beating? Knowing how poorly the blocks were made, we opted to move the cylinders and have some fun in the process.

As we rolled the five-foot-long, two-foot-diameter concrete cylinders, we *accidently* rolled them over rocks and watched them crumble. As Fidel said, they made crappy cement. We broke a lot more than we managed to move and got fired. Gosh, just when we had mastered the technique.

SIUTGA.

There was always something, no matter how grave the situation, that could make me laugh. When Fidel arrived at the camp one day, he left his car unattended. I watched as a guard, who probably had never seen anything more sophisticated than a bicycle or scooter, climbed aboard the vehicle, sit behind the steering wheel, and pretended to drive, turning the wheel from side to side, honking the horn, while two other guards rode in the back, bouncing along on an imaginary road. Don't know how far they got before they ran out of gas.

A Tragic Event

One of the more tragic events occurred while Fidel was at the Zoo. Earl Cobeil took a rather hard stance against Fidel and refused to comply with his demands, and Fidel wouldn't accept anything less than complete submission.

Fidel beat Cobeil so severely that he lost all reason and ability to think clearly. Fidel was relentless. When the beatings failed to have the desired results, Fidel had Cobeil hauled out of the cell one night and taken away. From the marks on Cobeil's hands and head when he returned, we guessed they had given him electric shock treatments.

Fidel then put the burden of keeping Cobeil alive on the men in his cell and told them that if Cobeil died, he would have them all killed. His cellmates tried helping Cobeil by force feeding him and caring for him in hopes of keeping him alive. But the last time Cobeil was taken from his cell, he was never seen again.

Vietnamese Mentality

During one of their many interrogation sessions, Bug showed me a copy of *Time* magazine and pointed to an RCA advertisement and asked, "Is this what you call television?"

There he was, a grown man with absolutely no idea what a television was. Even as a child growing up in New Mexico in 1956 I had seen a television. Although my family didn't own one, I remember going down the street to Stevie Brooks's house on Thursday evenings to watch the *Lone Ranger.*

It struck me just how wide a gulf of common experience there was between us. There we were about to sit down and have a rational discussion about the political and military conflicts between our two countries, and he had never seen a television. Lacking knowledge isn't a crime, but this man was toting a rifle.

Not only did he have no understanding of television, he had no concept of air conditioning. We were sitting in a room sweating like two lumberjacks in July, and I began telling him how in America we had entire cities built under huge domes cooled by refrigerated air. He couldn't quite comprehend the technology, but sat there in wide-eyed wonderment hanging onto my every word.

"And you know," I'd continue, "back in my country we have air conditioning inside our homes where we stay cool all day; and we climb into our air-conditioned cars and drive to air-conditioned offices."

Taking the fun further, I told him about going to a fancy, air-conditioned restaurant to eat dinner where we selected what we wanted from a menu listing hundreds of delicious entrees and desserts. I told him how a waiter came to the table and took our order and how we sipped wine until he brought our food and set it on a freshly ironed, white tablecloth. He sat bewildered with no concept of how that could be possible.

Then, I raised the bar. "We made love to tall, beautiful women on cool silk sheets, even taking on two at a time," I said. I didn't provide the graphic details. I just created pictures in his mind, and then sat back and watched his jaw-dropping reaction. He gobbled up everything I said like sweet cornbread.

Mindless Interrogations

Whenever a fellow prisoner returned from interrogation he would let other prisoners know what kind of interview it was: a serious right-to-the-gut, hard push interrogation; a Kiddie quiz (a session that amounted to nothing of substance); or just an attempt on their part to practice English.

Oftentimes when I was called in for a session, I'd notice a bulge under the blue tablecloth and a wire leading to a tape recorder under the table. This was a signal they wanted specific information and the session was being recorded and used as propaganda for soldiers in the South. Those were the serious sessions and were intended to garner information come hell or high water.

The camp officers knew I was a small fish, a mere lieutenant, and was not privy to high-level information that could be used to further their cause. When asked specific questions about strategic planning, I told them that if they wanted classified information they had to get it from the senior ranking officers, because I certainly didn't know what war plans were being strategized in Washington any more than the politicians back in Washington did. With the constant revision of war plans, I doubted any one knew the plan more than forty-eight hours ahead.

When Rabbit or Soft Soap Fairy, the head honchos of the prison camp, failed to obtain any vital information from me, they let a junior interrogator take over, and that's when the session became a mindless game, and I turned the session into fun.

The interrogators always began a question with the stock phrase, "So, how it is now that you . . ." It was as if they were reading from a prepared script.

One of their favorite questions was, "So, how it is now that you understand today the fortunes of the American war in Vietnam?"

I'd answer with, "I have come to understand that the so-called humane and lenient policy of the Vietnamese as applied toward the American prisoners is subject to international intentions." They heard the word "understand" and "humane and lenient" and were satisfied.

When they asked me what my name was and if I answered Edward George Mechenbier, they had an answer that included a critical element they recognized as something they wanted. It didn't matter that my middle name wasn't George, but John.

I was getting good at inventing a creative line of malarkey. During one session the interrogator wanted to convince me the Vietnamese could bomb San Francisco and had plans to do so. I asked for a piece of paper and a pen and then proceeded to give him a lecture on how far an MiG could fly without refueling. By the time I finished, he had a couple pages of rubbish that he probably used on other Americans later to show how smart he was.

It didn't take much to pull something over on these simpletons. Among the many lessons I learned at the Academy was, if I didn't know the answer to a question from *Contrails*, I could avoid a demerit by giving an answer that was interesting but not necessarily pertinent to the question. As long as I sounded authoritative I got away with it.

Most interrogators had only a limited grasp of the English language and no understanding of punctuation. It became a game to see how far I could push. If I sensed the session was intended only to practice English, I'd give them an off-the-wall answer, knowing it was no answer at all, and then sit back and watch their expression. They'd look at each other and then confer in Vietnamese. I would act as if I understood what they said, "No. No. That's not what I said."

That really bugged them.

"You understand Vietnamese?" they would ask.

"Maybe. I've been here a long time."

They would continue to talk among themselves and then ask me what was the word for horse?"

"That's *fermastam*," I would answer.

"What's a fermastam?"

"I can't think of the Vietnamese word that describes it."

Then they would offer a word.

"No. That's not it. Give me another one."

It was fun messing with their heads.

Rabbit had decent command of English, but many junior officers didn't know a verb from an adjective. It was hilarious listening to them as they slaughtered the language.

Spike Nasmyth, who wrote the book *2355 Days: A POW Story*, about his experiences as a POW, relates how he was once interrogated.

> "What you name?"
>
> "Nasmyth."
>
> "Nahshit, shit down."
>
> When Spike pretended not to understand the request, the situation turned into a comedy routine.
>
> "You must confess, or you will be punished."
>
> "I don't understand what you are talking about," he told the guard.
>
> "Nahshit, I will have you killed, and that could be very dangerous for you."
>
> "What?"
>
> "Nahshit, why you laugh? No laugh! No laugh! You must be punished. Keep shilent! Keep shilent! Kneel down!"
>
> "I don't understand."
>
> "Hand up! Hand up. Hand over head."
>
> "How can I put my hand over my head? I have a broken arm."
>
> "The other one. Queek! Queek! Keep hand up or I will allow guard to beat you. You have broken camp rules. You must be punished so you can think clearly."
>
> "Get up! Get up! Kneel! Kneel! Hand up!"

When I figured out that an interrogation was nothing more than an attempt to gain information about how Americans lived, I employed my evasion technique and redirected the questioning to benefit my purpose, which was to never give them anything of value.

I loved making it look as though we were the greatest pilots in the world, which we were, of course. When asked about the accuracy of our weapons I'd tell them how we could put a 20mm bullet between the eyes of a water buffalo at twenty thousand feet and never miss.

"But you are bombing churches, dams, dikes, old folks, women, and children," they'd say in protest.

"No, no. We're very good at this. We never miss. If we bombed something it's because we intended to."

Knowing that all they wanted was an answer to fulfill a requirement on a piece of paper, I gave them the old Mechenbier double-talk routine.

"That's a very interesting point because just the other day my cellmate and I were talking about the international situation relative to a lot of things is getting very interesting down to the basic material that is hard to comprehend. There are many things about a cow's stomach and because the heat generated when you look at the left side of the tongue of a blue-eyed cow . . . now did I tell you about the blue-eyed cow?"

I'd pause long enough to take a breath, and then continue, "You need to know this because they give higher butter fat milk, which is hard to transport in an airplane. When you start to do something like that you affect the center of gravity of the airplane. Next thing you know, the team made three outs in the bottom on the ninth inning and it was all over. I tell you it was all over! The victory was theirs. But what the communist did next was run a flag up the beanpole. And then the next day . . . "

I could have rambled on for hours. It didn't matter that what I said made no sense. As long as I was saying something, they were content. Especially if I talked fast and used big words, they sat there as if listening to some great philosopher. The situation would have been funny if it wasn't in Vietnam. I could almost feel a little pang of pity for the poor, ignorant, little creeps. Although I never knew what the Vietnamese were capable of, the one thing I never worried about was being brainwashed.

After a while the interrogations became so mundane that when they asked for the hundredth time, "So, how it is now that you understand today the fortunes of the American war in Vietnam?" all I had to do was nod and say I understood and the interview was over. They took that to mean I agreed with them. Nothing could have been further from the truth. But it worked and I stayed with it. They had an answer. It didn't matter that the answer made no sense.

Say What?

The Vietnamese had a talent for twisting news stories 180 degrees to their advantage.

One of the hot pieces of news we acquired during an interrogation was that the United States had landed a man on the moon. This was 1970, almost a year and a half after the event. One of the prisoners happened to see a partially burnt envelope in a trash pile with a postage stamp depicting the moon landing. He picked it up and asked the guard about it. The guard told him that he would soon learn.

A few days later the answer came via the *Voice of Vietnam*: "We have news showing the superior technical knowledge and capability and resources of the Soviet and Communist system. The Soviet Union put a highly sophisticated and technical unmanned vehicle on the surface of moon to report back much scientific data, while the United States lacking this technical capability could do no better than expose a man to the moon's dangerously harsh environment."

Another example of this came when an announcement was made over the loud speaker stating that in a recent dual track meet between the Soviet Union and the United States showing superiority of the socialist system, the Soviet Union came in second place while the United States could do no better than next to last.

Duh!

POW Facts

While at the Zoo, without the luxury of newspapers, radio, television, or magazines to keep our brains from shriveling to the size of a grape, when we had a question we'd tap on the wall confident one of the intelligent gentlemen living on the other side would have the answer to our queries.

We'd ask: What is the height of the Empire State Building? What is the longest river in the world? Each time hoping someone knew the answer. But the answer always came back, "I don't know."

Then beginning in the fall of 1969 we began getting answers. We wanted to know, how many home runs Mickey Mantle and Roger Maris hit their best year combined while playing for the New York Yankees? The answer shot back: "The year was 1951 when they hit 102 homers. Mantle hit 52, Maris hit 50."

We asked how many guys got promoted to major in 1966? The answer came back immediately, 1,622.

Being young, healthy fighter pilots we wanted to know who had the larger breasts, Marilyn Monroe or Jane Mansfield? Answer: "Monroe was a 38C while Mansfield was a voluptuous 42DD.

Wow! This was great. There must be some exceptionally smart guys living somewhere within the prison, and we were fortunate they gave of themselves tirelessly to address our endless list of questions.

Now, at Camp Faith, I asked Joe Milligan and Gary Anderson if they knew who the gurus were with all the answers to our questions back at the Zoo? "I want to thank them. It's refreshing after all those years of being told 'I don't know' to finally have authoritative answers," I said.

Gary got a big grin on his face and said, "That was me and Joe. We got tired of saying, 'I don't know.' We just made up all that crap."

Our entire body of knowledge gained through the prison walls was nothing more than a bunch of blarney. After that, any time we asked a question it was always prefaced by, "Is that for real or is that a POW Fact?"

The Son Tay Raid

On November 21, 1970, when the attempt to rescue American captives held at Son Tay failed and no prisoners were found, the North Vietnamese quickly changed their strategy and moved everyone back to Hanoi.

One chilly night in November I heard the sound of helicopters in the distance and noticed flashes of light reflecting off the ceiling of my cell. Looking out the window, I saw flares lighting up the horizon to the west followed by the impact of bombs shaking the ground. It wasn't unusual to hear explosions, but this far out in the countryside made me nervous and I wondered what was happening.

The next day we were confined to our cells while guards scurried around nervously.

Something was going on.

Two nights later, the eight of us were rounded up and told to hurry and gather our belongings; we were leaving. I was blindfolded, my hands tied, then loaded into the back of a vehicle. I had been moved so many times by now I had more miles than a Mayflower moving van.

CHAPTER 11

Camp Unity

CAMP UNITY, situated within the Hoa Lo prison, was made up of three long buildings with seven large rooms (bays) each able to house as many as thirty or forty men. Attached to each building were quiz rooms, and in the middle of the compound was a round building used for interrogations.

Two concrete slabs, raised eighteen inches from the concrete floor, ran the length of the rooms along two walls where men slept shoulder-to-shoulder. A double iron door stood at the end of each room with a small latrine at the other.

At the end of November 1970 I found myself back at the Hanoi Hilton once again. The night was extremely cold when we arrived, but the guards seemed unconcerned as they confiscated our blankets, leaving us shivering with only the clothes on our back. They took our cups, fans, mosquito nets, and even toothbrushes and toothpaste, looking for contraband.

Unlike before when I was housed with only one or two other men, now the Vietnamese put thirty-seven of us in a large room in a section of the prison called Camp Unity. Seeing so many Americans in one room at the same time was indescribable. And best of all, I was pleased to see Kevin among the group. I noticed he looked somewhat better than the last time I saw him, but not by much.

Another thing I noticed was some of the men, in place of the red and gray striped clothes like those we had been given in the earlier days of our captivity, were now wearing black pajamas, even their underwear was black.

There was an electrifying giddiness, bordering on euphoria, as we came face-to-face with guys we knew only by the sound of their tapping through the walls. I was astonished to find that the man I had communicated with for so long looked nothing like the man I had pictured in my mind.

Our laughter and commotion drowned out the guard's footsteps and key rattling as we whooped it up when we discovered a guy we had been stationed with many years ago was presently living just down the bed row. Comparing notes, I found that ten fellow Air Force Academy graduates from my class were also being held there. Guess you could say we were the class that got shot out of the sky.

I was delighted to see Don Spoon and Bill Baugh, the two guys I watched get shot down almost four years earlier. I had often wondered if they made it safely out of their aircraft.

We were on a natural high and I doubt anyone slept much that first night as we got reacquainted. In this new communal environment we were quick to identify who belonged to which military branch and joined in friendly camaraderie pitting each branch against the other. Most of us were either Air Force or Navy but we did have four Marines among us: Harley Chapman, Jerry Marvel, Larry Friese, and Orson Swindle—four really great guys.

Now confined in such large groups, the senior ranking officers whose duty, according to the Code of Conduct, was to assume command and maintain military discipline, decided we needed a chain of command as if we were on a military base in Somewhere, USA. Because we were part of the fourth war in which Americans were held on foreign soil, they created the 4th Allied POW Wing. Since there were three men from South Vietnam and four from Thailand among us, the term "Allied" aptly applied.

Feeling we needed structure, the SROs formed committees. One was a private communication committee for high-ranking officers only, which was silly if you think about it. In order to send information to these officers, messages had to go through the communication system consisting of lower ranking officers, and we ended up knowing everything the SROs knew anyway.

With little else to do, they dreamed up a set of rules called "Plums" that outlined how we were to conduct ourselves. One Plum explained how much torture we had to take before giving in. Other Plums instructed how to send and receive mail and under what circumstances we could accept a favor from the Vietnamese, such as medical treatment. One described what constituted a lawful escape attempt and under what conditions we should resist punishment. There were Plums that gave the order in which we were to be freed from prison, and how we were to conduct ourselves upon our release.

Some of their rules and regulations seemed hokey and excessive in light of our situation. Common sense served us well and we were already anchored in the Code of Conduct. Candidly, the Plums seemed to be a figment of the senior officer's imagined control, more inane than anything else.

Among the craziness the senior officers also expected us to commit to memory eighteen operating procedures. It was about this time that one of the more senior officers came up with the idea to send letters home to be forwarded to Washington, telling the Department of Defense that if the Americans were going to nuke Hanoi, they were to fly over and give us three sonic booms on three consecutive nights so we had time to prepare to die. We just shook our heads. I think they had too much free time on their hands.

Living with so many men jammed into a dormitory-style room could have been taxing, causing skirmishes and bouts of dissension, but that never happened. I never saw one guy become angry or push and shove another prisoner. We knew we were all in this mess together and needed each other to survive. We hung in there together to build a better quality of life within this ugly, austere, cold, and miserable existence.

Unity University

Although our bodies had been through hell, our minds were still sharp. We were all college educated and, coming from various backgrounds, collectively possessed a wealth of knowledge.

The men in each bay formed classes and turned our confinement into a full-fledged university with a dean who searched for men in their bay with special talents, skills, and expertise, and then formed a faculty. Soon students began lining up to register. The best part was we didn't have to take out college loans to pay tuition.

The curriculum rivaled any college in the United States. What it lacked in accuracy of fact, it made up for in enthusiasm, attendance, and shear lunacy. Classes ran all day, with oral exams and even a grading system. We kept classes short for Navy squids and Air Force bomber pukes with shorter attention spans and less mental agility. We jokingly assigned them tutors.

I chided the Navy guys, but they were truly a great bunch. Back in the spring of 1967 I had spent two weeks aboard the USS *Enterprise* and the USS *Kitty Hawk* with an F-4 squadron where we swapped stories and ideas on tactics and maneuvers. The assignment was more a boondoggle than anything else, but it gave me the opportunity to experience what it was like to be catapulted from a ship at sea in an airplane.

We took off and landed in a propeller-driven Grumman C-1 Trader called a COD (Carrier Onboard Deliver) where if you missed the tail hook you kept on rolling into the drink. For the Navy and Marines, flying off and onto a moving ship was routine, but for me, accustomed to landing on a ten-thousand-foot solid concrete runway, coming in for a landing on a three-hundred-foot metal landing strip that was heaving up and down, was intimidating and quite scary.

While returning from a training mission on the South China Sea, I was sitting in the back of the COD when the airplane suddenly went into a dive. This was not something you'd expect in a cargo aircraft. I looked out the window and saw flak all around us. It seems the pilot had drifted over Cat Hai Island off the coast of North Vietnam. Rather than flying out over the water and then into Da Nang, he missed the coast in point and got too close and the bad guys opened fire. The last thing I needed was to be shot down in a cargo airplane. How humiliating would that have been to see the headlines: "World's greatest Air Force fighter pilot shot down in the back end of a Navy C-1 airplane." I'd never live that down. Like I said, the Navy guys were a great bunch, but their navigational skills seemed lacking.

Anyway, back to the University.

Core studies consisted of political and social science, electrical engineering, electronics, international relations, geopolitics, geology, astronomy, astrophysics, foreign languages, and advanced math with algorithms. Electives included Automotive Mechanics for Dummies, Golfing in the Rough, Ship Building for Beginners, Celestial Navigation, Architecture, Animal Husbandry 101, Meat Cutting, Basic First Aid, Bee Keeping, and World History. We studied Beethoven, Mozart, and Chopin as part of the classical music workshop. To round out our education, we had seminars on classical and modern art, and studies on the works of Shakespeare, Byron, and Keats.

Read Mecleary taught sailing and explained how to navigate the high seas in a sailboat. I had never been much for water recreation, and doubted that I would ever have the opportunity to put the lessons to use, but I signed up anyway.

Danny Glenn, an architect, taught the importance of structural integrity in designing a house—how all the internal systems—electrical, plumbing, heating and air conditioning—must be integrated in order to work together. From his teachings I drew floor plans for my dream house.

Jack Tomes conducted Toastmaster sessions where we made speeches on topics of our choice. I heard speeches on everything from the manufacturing process of a paper cup to the proper way to eat crab legs. Some topics were enlightening, while others were just plain silly, but fun. Most of the men were gregarious and making a speech wasn't a problem. Getting them to shut up and sit down was another story.

Classes became the highlight of our day with men gathering in small groups to attend their scheduled class. Not having paper, pencils, or textbooks hampered our efforts, but we overcame this minor obstacle. By using pieces of whitewash broken off the walls as makeshift writing instruments we scribbled instructions and illustrations on the concrete floor.

Before long the Vietnamese gave us little blue notebooks, similar to college bluebooks, in which we were able to write our lessons. Perhaps there was a covert reason for this kindness. They could confiscate them to see if we were secretly making plans to revolt and overthrow the country—or maybe they were just curious about the American educational system.

Because we had no resource for researching information, when a question came up for which we didn't have the answer, we made up our own "credible reference" and reinstituted the "POW Fact" system established by Joe Milligan and Gary Anderson. After all who would know the difference?

Dave Luna, being Hispanic, taught Spanish. The first sentence Dave taught us was, "*Cuando yo vuelvo a mi casa en los Estados Unidos de América, voy a poner mi hueso del amor entre los piernas de mi esposa*" (When I return to my home, in the United States of America, I will put my bone of love between the legs of my wife).

Welcome to Spanish 101.

Jim Shively taught Russian. It took me almost a month to learn the Cyrillic alphabet before I began constructing sentences. Not an easy language to learn, that's for sure. Even a simple sentence looked strange. For example: *Здравствуйте. Меня зовут* translates to, Hello, my name is. . . . And to introduce your dog becomes, *это моя собака, пятно* (this is my dog, Spot). Like I said, learning Russian was a challenge.

I tried to conquer the Russian language, but had seen the writing on the wall, or the floor as it were—the Spanish students were having a lot more fun.

Kevin taught French. About the only lesson I took from his class was you should never speak French to a Frenchman because it infuriated them if you butchered their language. So, what's new? The French are always angry about something.

Being fluent in German, I taught the German class. My introductory sentence was, "*Können Sie mir sagen, wo der Bahnhof ist, Bitte?*" (Can you tell me where the train station is, please?). I used the KISS method of teaching, believing in keeping it simple.

A guard gave me a booklet bound in cloth in which I wrote a twenty-four-page German study guide. The guard routinely collected the book and inspected it for any sign of espionage. It was a hoot explaining the German language to the Vietnamese who barely understood English.

We had days when speaking English was forbidden. If a guy studying German needed to talk with someone taking Russian, for example, he had to find someone who could translate so he could talk with friends. Breaking the rules or lapsing into English, meant the offender was subjected to a kangaroo court and sentenced to a time out on his bed space for five or ten minutes.

While stationed in Europe, I had learned about the wines of various regions in France such as Alsace, Champagne, Loire, the Bordeaux and Burgundy areas, and the Rhône Valley. My tutor was the sommelier (okay, so he was the wine steward at the RAF Bentwaters Officers Club, but he knew a lot about wines). He taught me the subtle nuances of the French wines, their geography, and the way the soil impacted the wine's taste and quality.

Being a fast study and not above imbibing from time to time, I picked up the information quickly and as the self-appointed wine connoisseur, conducted a wine tasting class for my fellow inmates.

"Image that you are having a nice dinner and you begin with a cheese appetizer such as brie," I explained. "You might choose a very light Beaujolais or if you're serving a more pungent cheese such as Camembert, you'd want to try a fruity Chenin Blanc." I instructed the men to collect their tin cups and I "poured" them each a sampling.

"Before you swig that down, you must first taste the wine with your eyes. Hold your glass to the light and gently swirl the wine around. Does it cling to the glass? Does it have legs?"

One man nodded. One fellow had his cup upside down as if looking for the legs.

I pressed onward.

"Next, smell the bouquet." I waved the cup under my nose and watched as each man sniffed as though the finest of wines was in his cup. Since taste is reportedly 90 percent a matter of aroma versus actual contact with the taste buds, this required imagination.

"Now, for the taste." I pretended to take a sip and swirl the nectar around in my mouth.

The students followed suit. By their expressions, I could see they were really enjoying this. One guy licked his lips; another looked wistfully at his cup.

"The next step is to cleanse the palate. Take a bite of bread and a sip of water."

"Bread and water. Now, that I'm familiar with," a student piped up.

I ignored the comment.

"For the entrée, you are serving lamb. For that dish try a light-bodied Pinot Noir from the Burgundy region. But if you're serving pasta with a red sauce, try Cabernet Sauvignon."

"Well, which is it, lamb or spaghetti? Frankly, I'd rather have a hot dog," someone called from back of the room.

"Frankly, my dear, I don't give a damn," another man butted in.

"I'd settle for a Coke and a peanut butter and jelly sandwich," contributed another.

Trying to educate this bunch of rowdies was becoming a challenge. Their attention span was waning. I needed to pump it up. When they settled down, I continued.

"Cabernet Sauvignon should be served in a large bowl, to provide ample surface area for the wine's aroma to breathe. Don't fill the glass too full. Pour the wine only half way to the top to allow the wine to interact with the air. This will help it develop a full flavor for your nose and mouth." I paused giving them time to refill their glass.

"Now, for dessert. You've decided to have a chocolate mousse, you definitely want a Merlot with its sweet but soft plum flavor."

"I've had enough of Plums, thank you," someone said.

I knew I was losing them. Before they grew too restless I finished up with a Salud, Prosit, and Cheers, and then called it a night.

The Music Man

Al Stafford had somehow acquired a guitar from the guards. He was an accomplished musician and played classical guitar as well as jazz riffs. The music was a treat and a break in the monotony of an otherwise dull day.

One afternoon, he began strumming a tune. I found a piece of tile and by scraping it against the concrete, provided the percussion. We got pretty darn good. I enjoyed the interlude so much that he taught me a few chords and let me play the instrument.

The Vietnamese allowed him keep the guitar during the day but took it back at night. This went on for several weeks until they took away the guitar permanently. Nothing good lasted very long in prison.

Movie Night

Mondays became movie night—sans popcorn, Cokes, and Milk Duds, as John Fer, a 1962 Academy graduate, recreated scenes from popular movies. We sat listening as he reenacted movies so vividly it was as though we were sitting in a theater watching the movie flicker on the big silver screen, except here in our cell they were shown on the concrete walls from an imaginary projection room.

John had the talent to walk and talk like John Wayne as he acted out scenes from his favorite films. It was amusing to watch him swagger across the "stage" in his affected John Wayne drawl saying, "That'll be the day!" from the movie the *Searchers*, or "Don't apologize—it's a sign of weakness" from *She Wore a Yellow Ribbon.* He was a show unto himself.

After retelling the movies a hundred times, we began memorizing the dialog and soon took on the role of characters in the movie and acted out scenes. Some movies such as *How the West was Won*, a long movie to begin with, lasted days with all the embellishments we added. In our own goofy way, we created a fantasy world for a few hours a day.

Pinkeye

We believed in sharing. When one guy got pinkeye, we all got it.

I was careful not to put my hands near my eyes or touch anything someone else had handled. Despite my best efforts, I still contracted the highly contagious disease. As if I didn't have enough problems with my back and teeth, now my eyes were red, puffy and swollen and I could barely see. I had to have help getting my food and finding my latrine pot. With everyone having the same problem, it was like the blind leading the blind. The Vietnamese, of course, did nothing for us but wash our eyes out with a dirty water solution.

Mike Burns, notorious for getting pinkeye, managed to catch it several times and became an invalid for months. Unable to see and pull his share of the workload, we had to help him fold his blanket and do his daily chores, but after contracting pinkeye for the third time and seeing how he was milking his condition for all it was worth, we found a rock, consecrated it a brick, and presented him the Gold Brick Award.

Rolling Cigarettes

With the excuse that the Americans had bombed their cigarette factory, we often didn't get our customary daily rations of cigarettes. Instead they gave us bulk tobacco and told us to make our own. Pinkeye left most men unable to see, so when I wasn't afflicted, I got the job of rolling cigarettes for those less able.

Unlike cowboys in old western movies who sprinkled tobacco onto a piece of paper, licked the edge, and then rolled it between their fingers, I learned to roll a cigarette Vietnamese style. A guard gave me papers and bulk tobacco, and then showed me how to wrap the paper around a stick before adding the tobacco. Rolling a nice tight cigarette is a skill set that takes a steady hand, and being a fighter pilot I had no problem with the task. I sat in the corner and for several hours a day rolled cigarettes for the men in my cell.

Although I wasn't a real smoker, when Ken Cordier, a graying but energetic gentleman, received a pack of Borkum Riff pipe tobacco from home, I rolled a cigarette for myself and smoked it. I couldn't handle the raw Vietnam tobacco, which burned too harsh, but the Borkum Riff was smooth and flavorful, and I enjoyed smoking it on occasion.

The Banana Peel Duel

In celebration of the anniversary of the Vietnamese revolution, the guards gave us a chess set. The pieces were nothing fancy, just an ordinary classic style set.

I challenged Tom Hall, my bunk neighbor, to a friendly match. Tom's F-8 Crusader had been hit, but he managed to limp his airplane back to the Gulf of Tonkin where he was picked up by a rescue helicopter only to "wreck" another one of Uncle Sam's multimillion dollar flying machines two days later and ended up with the rest of us. He told us how just before taking the stroke on the catapult, he made the traditional salute to the launch officer with his left hand indicating he, the pilot, was ready to go, but not to be confused with the regulars, he saluted with his middle finger touching the FUBIJAR stenciled across his helmet. What else can you say about a guy like that?

The chess match between Tom and me was to be a contest of conquering warriors pitting their generalship against one another. We dressed for the event—shirts, shorts, sandals, ties not required—and sat on the concrete slab that served as our bed and began the challenge. The game was totally unimportant except that I won. Tom rechallenged me and won. I took the third match, but he triumphed in the fourth.

The matches had become a source of entertainment for the rest of the men in the room, and they waited expectantly for the final showdown, the tie-breaker. Unfortunately, the guards confiscated the chess set in one of their regular demonstrations of superiority and control, leaving the matter of honor unresolved.

With the winner still undetermined, tension simmered for days waiting for the right moment to challenge my opponent again in some other contest. That moment came one morning as the gong sounded reveille. Our sleeping accommodations left us with little room to move and as I bent to fold my mat, Tom and I bumped butts.

Oh, the indignity!

"You, sir, have offended my honor by invading my space," I said in mock indignation and feigned a slap across his cheek.

Tom, who had assumed the title of Count Renwick from the Isle of Wight to enhance his stature among those of us who never had the privilege of visiting Tidewater, Virginia, was a tall, lean, baby-faced kid with a charming wit that runs in perhaps one in ten thousand human beings. He returned the mock slap, and once again the challenge was on.

As the resident German instructor, I rallied students from the German and French class to my side. I used our lesson time to defame, malign, and expose the crude nature of my opponent by scratching disparaging remarks on the floor.

Tom, a student in the Russian course, encouraged support for his side by organizing the Spanish and Russian students into alliance with him. With the French and German students pitted against the Spanish and Russian students, the feud continued for two weeks with charge and countercharge scrawled on the concrete floor.

Finally, Al Stafford, our SRO, announced that he had had enough of our posturing and declared that the matter of honor be resolved by a duel. Since our diet, on rare occasions, provided each man with one banana, we chose bananas peels as weapons.

The Vietnamese were always a bit edgy, and we were concerned that if we started a full-scale banana peel fight they would interpret it as a riot. Conducting a duel would be a gross violation of camp rules, so we had to be extra careful not to get caught engaged in such folly. We had to make our first shot our best and only shot.

To ensure the rules were followed, we needed a judge whose decision would be binding. We agreed he must be a man of unquestionable character, with a sense of humor, and above all, be older and held in high esteem. We promptly agreed that Billy Means, a B-66 Destroyer bomber puke, was the man for the job. Billy was a wonderful gentleman with elfish characteristics, pointy ears and thin features that earned him the nickname, "Pixie."

We decided the best time for the duel would be after the second meal in the day while most of the guards were eating, giving us a better chance of settling the feud without interruption. With the day and time set, we waited for the appointed hour.

"Gentleman, your attention please," said the judge. "Introducing the contestant on my right, the Morgantown Moon Beam."

I had rightly earned the name, having been born in Morgantown, West Virginia. The Moon Beam surname personified my purity and honesty.

Whispered cheers erupted from my side of the room as muttered jeers and sneers came from my opponent's corner.

I strolled to the center of the room dressed in white socks, t-shirt and shorts, and a white hat fashioned from a handkerchief—items the Vietnamese allowed me to keep from my Christmas packages as part of their humane and lenient policy toward the American Air Pirates. Several small white handkerchiefs tied together and wrapped around my shoulders for a cape, completed the ensemble.

Charles Stackhouse, a relatively rotund gentleman considering our eight-hundred-calorie-a-day diet, stood at my left side and dutifully one step behind, acting as my Second. Charlie was a super fellow and trusted by all. Similarly clad, he carried my weapon—a fresh, unmarked, beautiful banana peel of Chiquita quality on a blanket rolled into a pillow and covered by a white cloth.

"And on my left, Count Renwick from the Isle of Wight," announced Judge Means.

Tom's loyal followers gave a high-spirited hurrah as from out of the shadows slinked my sleazy opponent dressed totally in black. His evil looking Second, Ben Ringsdorf, a scrawny six-foot version of Bela Lugosi and Snidely Whiplash, lagged behind carrying his master's weapon, the blackest, most rancid peel imaginable, one they had kept hidden for several days.

We approached the judge who examined our weapons, inspecting each banana peel for texture, weight, aerodynamic stability, and size.

All during this preliminary ceremony my opponent hissed and scratched at me while muttering profanities appropriate to his character. I, the Morgantown Moon Beam, on the other hand, was totally unaffected by these crudities and prayed for my opponent's demented, tortured soul.

"Gentlemen, this is a duel to settle a matter of grave personal honor. Standing back to back, you are to take four steps forward, one step at a time on my count. When each man has taken his paces, he will turn and fire. Do you understand?"

We nodded in unison, all the while thinking of a devious way to outwit our opponent.

"Gentleman, take your places back to back. Ready . . . one."

"Yea, though I walk through the valley of the shadow of death, I shall fear no evil . . ." I said.

"You dirty miserable rotten SOB. I'm gonna shove this banana. . . ." Tom said.

"Two!"

"Thy rod and thy staff, they comfort me . . ."

"Hiss. Boo," my opponent's cohorts sneered.

"You miserable rat. You're going down. You're gonna die!" Tom said.

"Surely, goodness and mercy shall follow me . . ."

"Three!"

Not waiting for the count of four, and overcoming the fear of suffering a loss of honor, and as preplanned, we whirled around, drew back our banana peel, took careful aim, and fired them at Billy Means.

Pandemonium erupted. Everyone hooted and hollered.

Hearing the ruckus, guards assumed we were rioting and rushed in with guns drawn to squelch the uprising. They locked us down and took Tom and me out for questioning.

"Why you pick on little guy?" they asked.

"Because he's a bomber puke and we don't like him. That's why."

We could barely keep from laughing. The guards, however, seemed to frown upon such levity.

Sunday Church Service

Religion did play an important role in the lives of many POWs. While most prisoners believed that without their belief in God they would never have survived, not everyone had an epiphany or "found" Jesus during their imprisonment.

I was raised in a strict Catholic family and my faith was essential in keeping me grounded. I never came to a kumbaya moment that God was protecting me, but had a confident realization that He was watching over me, giving me strength to overcome and endure whatever the Vietnamese dished out. Because of my faith, I never lost hope that one day I would be released.

For some, there was a quiet reverence knowing we had been spared. We had survived the trauma of ejecting from an airplane, some in better or worse condition, but at least we hadn't been killed. Most of us felt our survival didn't happen because of any skill or cunning on our part, but was the providential entity of God that said there was a reason we didn't die when we were shot down. I think most of us had that attitude. We didn't need to walk around making the sign of the cross or praying every hour; it was something we felt inside.

In the earlier days, being isolated with only two men to a cell, we prayed quietly in our individual cells. Robbie Risner, our SRO at the time, instructed that each new "shoot down" was to be told that every Sunday at noon when the gong sounded for the guard's siesta we were to conduct a church service. Together, but separated by walls, and in keeping with the ecumenical spirit, we faced east, toward America, and recited the Lord's Prayer, the Psalm 23, and the Pledge of Allegiance.

Now, at Unity, with thirty men gathered in one cell, we held a more formal observance. Every Sunday at noon, each room held chapel call with a chaplain who conducted a short service. There were no fire and brimstone speeches or sleepy sermons, just a few minutes of common prayer and reflection.

We shared the honor of conducting the service by rotating duties every month. As a boy, I had memorized the words to the Consecration, which is generic to the Episcopal, Lutheran, and Catholic Church, and added that to the agenda when it came my turn.

As before, we recited the Lord's Prayer, the Psalm 23, and the Pledge of Allegiance, but now we added hymns. Any noise in the camp was strictly forbidden and we were pushing the limit by breaking this cardinal rule by singing. Despite the camp commander's edict, we sang the "The Star-Spangled Banner," "Amazing Grace," and the "Battle Hymn of the Republic." Imagine three hundred voices raised in song, reverberating though the walls of the prison camp. We must have sounded awesome.

One particular Sunday, hearing us singing, guards rushed in and pulled out the leader in each room and took him to a torture room for a little "physical atonement." Having little freedom to begin with, we weren't going to stand by and allow our religious freedom to be trampled. The next man in line stood and continued the service.

After two more leaders were removed, it was my turn to lead the service. I continued the service, and like the others, I was taken out of the room for correction. Rabbit, the camp commander, stated that we had embarrassed him in front of his Russian allies and we were forbidden further church services. If we were inclined to continue such activity we would be punished severely, even to the point of death.

I was shaking in my sandals standing before Rabbit, but I was also furious to think he could deny us the opportunity to practice our faith and persecute us like the Romans did the Christians.

With testicular fortitude, I stood up to him.

Guy Gruters, a fellow cellmate, related later what he heard through the walls as I told the guards, "You can torture and kill every one of us if you want, but as long as one man is still alive we will have our church service." I'm grateful that Guy remembered the incident because I was too scared to think, much less recall what I had said, but I do remember looking Rabbit in the eyes and saying, "Are you afraid of God?"

Rabbit looked startled and appeared physically taken back by my comment. How could I, or anyone, accuse him of something so heinous as to question his belief in God? It was as though he realized he had impugned a higher power.

I expected to be hauled off, beaten, and thrown into solitary, but surprisingly he not only backed down, he allowed us to continue our Sunday services, and come Christmas gave us Bibles. We had asked for Bibles for years only to be denied.

Each room was allowed to have the Bibles for only a few hours on Christmas Eve and Christmas day, but even that little time was joyous. Using crudely made pens and ink, men in each room made copies of as many passages as they could in the limited time we had the Bibles. To avoid duplications, each man copied different verses, and then passed them through the communication system to other rooms where they were memorized and then destroyed. Having a guard find the writing would have invited trouble.

Camp rules stated there should be no more than four men in a group at any one time. As Lent came around in 1971, the thirteen Catholics in the room thumbed our noses at this rule and formed a tight circle in the far corner of the cell away from the door to say the rosary. The non-Catholics stood in the middle of the room creating a shield should guards come nosing around. Before long our little group grew until the only men left standing as lookouts were the two Jews among us who kept a careful watch for guards.

The Virtual Christmas Gift

It was hard emotionally for many men to be away from loved ones at Christmas time. The memories of singing Christmas carols, gaily wrapped packages, kissing under the mistletoe, the smiles on children's faces as they anticipated the coming of Santa Claus, and Christmas Eve spent assembling toys that seemed to always come without instructions, brought a quiet introspection among the men. But Christmas of 1970 we weren't going to sit around feeling melancholy. We refused to allow a somber mood to overshadow the spirit of the season. We decided to exchange gifts.

Without the means to shop and purchase real presents, we created a virtual Christmas gift exchange. We set our imaginations free to shop and select the right present for the man whose name we had drawn from the hat, or tin cup, if you will.

I drew the name of Galand Kramer, a man with an astonishing memory who could recite the names of more than 750 baseball players. We had spent many hours playing Old Maid and Black Jack with a deck of cards we fashioned from bits of paper. Galand loved Baby Ruth candy bars, and being a world-class card player, he almost always won. I ended up owing him 244 bars of chocolate. He must have cheated.

Selecting his gift was easy. All I had to do was "drive" to the local candy store and "buy" his favorite treat. Galand seemed delighted when I described the colorful holly-imprinted red and green paper I had carefully chosen for wrapping his special gift.

Tim Sullivan of Boston drew my name. I had told him about my modest collection of beer mugs, purchased while traveling around the world. Come time to "open" my gift, Tim described the beautiful pewter mug he had found in a local shop and the special message he had engraved on it.

Our stay at Camp Unity lasted six more months before the Vietnamese decided we were having way too much fun and becoming too well organized, so they began separating us and farming us out to other camps.

CHAPTER 12

Zoo Library

In May 1971 I was moved again and found myself back at the Zoo. This time I was placed in the Library with Ben Ringsdorf and Tom Hall, both from the Banana Peel Duel; Ron Bliss, the boozer; and Wayne Smith, the all-American boy. Here we were afforded the freedom to be out of our cells and visit with other prisoners for a few minutes each day.

Dentistry Vietnamese Style

After years of neglect, the tooth that had been loosened by a guard's well-placed kick had become abscessed and extremely painful. I complained about the stabbing pain, but the camp officer did nothing until the left side of my face began to swell like a balloon.

After repeated requests for a painkiller or treatment, he consented to send for the doctor to "fix tooth." Doctor Zorba, who acquired the label because he looked like the character on the *Ben Casey* TV show with fuzzy hair sticking out from under his white surgical cap, appeared ready to perform his dentistry skills.

I was told to sit in a chair while two soldiers held my shoulders and two others held my legs. Zorba then took a Novocain-filled needle and jabbed it into my gum. He seemed to derive great delight in stabbing the needle everywhere but where it needed to go. I muttered and pointed to the maxillary gland in the back of my mouth. After a few more attempts, he finally found the correct spot.

Rather than wait for the Novocain to numb the nerve, he took a chrome-plated chisel and began tapping around the tooth. (I'm guessing to verify that he had the correct tooth, as if my grunts of discomfort weren't a clue.) A sharp pain shot up through my skull with each tap. Then he took a pair of pliers, and without hesitation, yanked out the tooth. It wasn't one of those small teeth near the front of my mouth; it was one of the big molars in back. I shot up a foot off the chair, screaming in pain, and would have hurt myself, or the dentist, had the soldiers not been holding me down.

Proud of his accomplishment, Zorba held up the tooth and said, "Now, tooth fixed."

That evening a guard gave me a cup of warm powdered milk. Considering the milk a special favor, I refused to take it, but Al Stafford instructed me to drink it because I hadn't eaten anything solid for a long time and needed the nourishment. I took a sip and passed the milk around to my cellmates, but the guard caught me and that was the end of the milk.

Even after the pain subsided and I began to feel better, the hole left by the extraction continued to drain and I spit blood for several days.

Utopian Potty Seat

Did I mention that we were the world's greatest fighter pilots? We were smart, well-educated geniuses. There wasn't anything we didn't know. Our intellect rivaled that of Einstein and Edison.

For many years, we did our personal business in the corner of the room while sitting on our pots, hoping it smelled like Essence of Rose Petals. Most of our latrine buckets were small, rusty, sharp-edged affairs, and sitting on them often left a sore, red line of demarcation on our backsides.

One afternoon, Tom Hall was sitting on his bucket being observed by none other than Ben Ringsdorf when Ben got a brilliant idea—pilots do that—and said, "You know, Tom, it would be a lot more comfortable if you placed your rubber sandals on the rim of that rusty bucket before you sat down."

Well, Duh!

And with that the Utopian Potty, an invention worthy of a patent, was born.

Letters From Home

In January some forty-two months after my arrival in Hanoi, I got my first letter from home, and the lies I had told years earlier came back to bite me in the ass.

Prison authorities had opened and read my letter from Jerri and found that some of the information I had given them didn't match what was in their blue book.

"How many brothers and sisters you have?"

I couldn't remember what I had told them.

"Who is Jan?" the interrogator asked.

I gave them a blank stare hoping my expression wouldn't reveal any acknowledgement.

"Who is Lora? Who is Chuck?"

I shrugged.

"Who is Joellen?"

Although Jan, Lora, and Chuck were my siblings, I didn't have a sister named Joellen. I figured they were pitching out names trying to get a reaction from me as part of their game.

When they pressed for an answer, I told them that my mother was pregnant with me when my father was killed in the Second World War. She remarried when I was much older, so I had a half sister who I barely knew. That's probably who she was, I told them. They must have bought my load of bull, because they eventually gave me Jerri's letter.

Many men cherished such letters, but I read Jerri's without any emotional attachment. I looked at it only as a way of collecting information I could pass to others. Like so many men, I had shielded myself from feeling any sentiment. We felt everything happening to us wasn't real and believed we would be going home soon. I packed all feelings away in the back of my mind and did not allow them to surface with a false face.

I read the letter and gave it back. It was safer that way.

Forming a Brotherhood

The Vietnamese not only enjoyed taunting us by playing songs over the loud speaker that contained subtle messages, they also enjoyed trying to break our morale by telling us our wives were being unfaithful, had abandoned us, and had remarried.

Soft Soap Fairy twisted a comment Jerri made in her letter about a pink jeep with the fringe on top.

"Your wife, she is going to movies outside in pink jeep with other men. Your wife, she is being unfaithful to you," he said.

Before I left for Vietnam, Jerri and I had talked about our next assignment to Duluth, Minnesota, after my tour in Vietnam. We had decided we'd need a jeep to plow though the snow, and called it our surrey with fringe on top. In her letter to me she must have referred to that jeep.

I made a flippant remark that if my wife was going to an outdoor movie with other men on a cold January night in an open jeep, she deserved to freeze.

"But she is being unfaithful. She has abandoned you," Soft Soap Fairy replied, seemingly pleased that he could make such an astute statement.

"You shouldn't judge American women by Vietnamese standards," I said.

That remark crossed the line. The guard backhanded me. My face stung as though he had slapped me with a hot steam iron. Then he proceeded to work me over, kicking me in my ribs and crotch until I couldn't stand.

Doubled over in pain, I shuffled back to my cell with a guard following close behind. Seeing me in that condition, the men wanted to know what had happened. I told them what the guard had said about our wives and my comment not to judge American women by Vietnam standards.

Later, I heard that at least three other men used that same line while being interrogated. Their comments not only took the heat off me, but also let the Vietnamese know that we stood together as brothers. The fact that those terrific guys gave no thought to their own safety and comfort and in essence said "Up yours" to the Vietnamese showed our solidarity and meant a lot to me. That's the kind of bond we had. We stuck together no matter the consequences.

Taking Care of Each Other

Just as our emotional survival depended on a sense of humor, our physical survival hinged on taking care of each other when we were unable to help ourselves. I've already told you about popping Kevin's boils, but there were many other incidences where we had to rely on each other.

Galand Kramer, who had been my cellmate back in Garage 4, had extremely bad asthma. Sometimes at night his condition became so acute that Ron Bliss, Kevin, and I took turns sleeping with our ear next to his face, listening in case he stopped breathing, which he often did. We'd pound on the door and yell for Boxi who came and gave him a shot. We never knew what the shot was but it worked.

My tooth wasn't my only problem. The cold weather bothered the knee I had injured while playing intramural football at the Academy; and with my back injured during the ejection from my aircraft, there were times I could barely walk. The polluted air and the mold in my cell stirred up allergies I had developed as a kid and led to terrible sneezing fits. My cellmates had to hold me down to keep me from further aggravating my back.

I also had a very dry skin condition on my upper back that I couldn't reach to scratch. Again, Tim came to my rescue and volunteered to scrub my back with lye soap and a rag to exfoliate the dry, dead skin, in an effort to stimulate the blood flow and relieve the constant itching. Without helping and caring for each other, it is doubtful that many of us would have survived as well as we did.

The Picture Under the Mat

Herv Stockman wasn't the only talented artist in the group. I'm not in the same league with Stockman, but I can draw fairly well. One afternoon when we were allowed to mingle with men in the cell next to us, I used the opportunity to slip into Jim Shively's cell and draw a picture of a full-length, rather shapely, seminude woman on the concrete slab under Jim's mat.

When the guards came to inspect his cell, they lifted the edge of the mat and saw the picture. They stood goggling over the drawing for quite some time, elbow jabbing each other, snickering, and taking a bit longer than usual to search the room. It gladdened my heart to give them their very own version of *Playboy.*

CHAPTER 13

Dogpatch

ONE HUNDRED AND FIVE MILES north of Hanoi on the remote Chinese border in Cao Bang Province, sits the small compound the Americans called Dogpatch. A brick wall covered with barbed wire and camouflaged with foliage surrounded sixteen buildings constructed of stone and concrete.

It was a dark and damp place with dungeon-like cells, slits for windows, and thick wooden doors. Situated between two mountain ranges, sunlight reaching the camp was limited to only a few hours each day.

The Trip North with Boo Boo

Late one night in May 1972 I heard the rattling of keys followed by a guard entering my cell. He told Ben, Tom, Ron, and me to gather our belongings. We were being moved again.

Handcuffed and blindfolded, we were taken two by two and crammed into the back of a truck, part of a sixteen-truck convoy. Along with twenty other prisoners, we sat shackled together with our legs bent to our chest, like a bunch of kids trying to set a world record by seeing how many bodies could fit inside a Volkswagen Beatle.

At first I thought perhaps we were going back to Camp Faith, but the hour-long drive turned into two hours, and then three as we traveled up into the mountains on narrow, bumpy roads. The trip lasted until the evening of the next day, and I began to realize this was no ordinary move and feared we were being taken across the border into China, a country from which I felt certain we would not return.

Perhaps one reason the trip took so long was the driver didn't know where he was going. We knew they had gotten lost a couple times when the driver stopped, asked directions, and then backtracked.

Another reason for the long ride was because the guards stopped at every village. We could hear them talking to villagers as they flipped back the canvas cover to show us off, proving how important they were as emissaries of their government in charge of American criminals. But I'm just guessing.

From time to time, they pulled off the road for gas and allow the guards a latrine break, but we were never permitted out of the truck—we had to hold it. Except for a little water, they gave us nothing of substance during the entire trip.

Boo Boo, the happy guard, was small in stature and spoke very little English. Having the duty of being our guard on the trip, he sat quietly in the back of the truck holding his rifle in one hand and the bullet clip in the other.

At one point, when the truck stopped, and needing to get out, Boo Boo realized he couldn't reach the ground while holding onto his rifle and bullets at the same time. You could almost see his mind's gears grinding as he looked from the gun and then to the bullets, pondering his predicament. Finally, in resignation, he handed his gun to one prisoner and the bullets to another, and then gestured for the two men to stay apart. After climbing down he retrieved the weapon and ammunition. Problem solved.

After thirty-six hours jostling along bumpy roads, we arrived at a camp called Dogpatch. It was dark as I climbed from the back of the vehicle. My back ached and my legs, held in one position for so long, were stiff and sore. When I tried standing, they buckled under me.

Cobras in the Bed

Bob Wideman, yes, the snorer, and I were assigned as cellmates once again. We were getting settled in when I heard a gunshot. Gunfire was common, so I didn't think much of it.

Unlike other camps where there was always a bare light bulb hanging from the ceiling, here there was no electricity and we spent that first night in total darkness. It wasn't until morning that I saw the condition of our cell. About six by seven feet, Bob and I had to sleep on an L shape concrete slab with our feet literally touching.

I also found out what the gunshot was all about. The story goes that the guard, after assigning two men to a cell that hadn't been used in quite some time, left them and indicated he would return shortly with a lamp. The men sat on the concrete slabs, holding their bedding and belongings, waiting for the guard's return.

Some time later, the guard returned carrying a lamp and opened the door to find a cobra, head reared, sitting between the two men as if posing for a class photo. The guard blasted the cobra with one shot. I can't testify as to what the two men must have thought at that moment but I'm guessing a few extra prayers were said that night.

Upon further inspection, the guards found that cobras had nested under the building. We dubbed that building "Cobra" and kept our distance.

We had had snakes back at Zoo, but they only came around during the rainy season. Here in the mountains seeing a snake was an everyday occurrence. I tried to keep my distance and never walked into my dark cell without making noise before settling in. It wasn't uncommon for a man to wake up and find a snake curled up in his bed.

Dogpatch was different from other camps where I had been held. There was a small common area outside our cell enclosed by a high wall. Although we were locked in small, two-man cells at night, the doors were opened during the day, allowing men in adjoining cells semifree rein to walk about and mingle with other POWs. The treatment was somewhat less threatening, but the food remained the same. I saw Needles once in a while, but we never received any medical treatment. We had no running water, and the only time I got to bathe was after a rain when we enjoyed a quasi-shower by standing under sun-warmed rainwater running off the metal roof.

Moonshine Makers

We soon discovered we had a rumrunner among us. It's probably best not to mention Tom Hall's name as the mastermind, but he told us that if we took the bananas and orange-like fruit they gave us, put them in a bowl, and allowed the mixture to ferment, we could make some pretty good homemade hooch.

As directed, Bob and I took the rinds from the oranges and pieces of banana, put them in a bowl, covered it with a piece of damp cloth, and let it sit a few weeks. We kept the bowl hidden under the bed and out of sight of guards.

Each day we watched in anticipation as the cloth ballooned and the juices bubbled and brewed. When the mixture didn't turn green or moldy, we knew we were on to something great. The elixir was just days away from being ready and we eagerly awaited our first taste. But before we could imbibe, the guards, like mountain revenuers, pounced on our still and closed us down. That was the end of our moonshine making business. Too bad, we were looking forward to a good buzz.

Stitching a Picture of Jerri

The Vietnamese never allowed us to keep photographs from home and the picture Jerri had sent me was confiscated soon after I saw it. Holding that picture in my mind, I found a piece of insulation wire, stripped it bare, and then bent the end to form an eye and made a needle. Pulling pieces of threat from my blanket, I carefully stitched a picture of my wife onto my two-tone red and grey, short sleeve shirt.

The picture wasn't the quality of a Gainsborough, but it resembled a woman sitting with her legs curled under her.

Most of us suffered from parasites. Dave Luna suggested a good way to get rid of them was to eat the small, hot red peppers that grew wild around the camp. We picked them and ate small, and I do mean very small amounts, at our own risk. The peppers burned going down, but did the job ridding us of any vermin living inside us. We figured if the peppers didn't kill us and got rid of the parasites, we'd be okay.

Utopian Lamp Contest

With the camp situated in a ravine between two mountains that prevented sunlight from reaching us except from 10:00 a.m. to 2:00 p.m., and with no electricity in our cells, we spent many hours sitting in darkness. The Vietnamese gave us small cans, similar to tuna or cat food cans, and kerosene oil to make lamps. By putting a little oil in the bottom of the can and attaching a wick to a piece of tile placed in the bottom, presto we had light, albeit ever so dim.

Ben Ringsdorf, a most talented gentleman, found that by bending the tin he could control the wick thus controlling the size of the flame, making the oil last through the night. That led Ben to discover that if he also twisted the tin he could make interesting shapes and designs. Always looking for something to do, someone came up with the bright idea of having a contest to see who could create the most unusual lamp.

The challenge was on, and thus began the Utopian Lamp Contest.

The ten men in our building worked eagerly designing unique creations. By bending, contorting, and deforming the tin we fashioned handles and legs for our lamps. My simple work of art couldn't compete with the piece made by the very talented Tom Hall, my banana duel conspirator. He twisted the metal into unusual shapes, making a lamp with filigree handles and even created a device for hanging it, winning the contest hands down.

Setting New Priorities

When we first became prisoners, we had well-established priorities. We liked certain foods, music, books, and women. But as days grew to months, then years, we came to understand our hierarchy of needs had changed. What we missed and desired most and what we once considered critical to our every day lifestyle no longer held the same value and had become less important.

Taking a survey we agreed that the food we dreamed of most that first year was steak and lobster. The second year it was hamburger and hot dogs. After four years our palate would settle for a peanut butter and jelly sandwich.

The question posed to the group was: "If you could choose any woman in the world, to be locked up with who would it be?" We could have voted for the sexy Bridgette Bardo, the buxom Jane Mansfield, or the intriguing Zsa Zsa Gabor to fulfill our fantasy, but after our long stint behind bars, locked up like animals, we selected Mary Tyler Moore. Surprisingly, we had acquired a greater understanding of what was important and determined that, although sex was great, our values had become more realistic. Homespun won over glitzy sex objects.

In January 1973,a guard rushed into our cell and directed Bob and me to roll up our blankets, mats and nets, gather our belongings, and prepare to move out. There was a flurry of activity in the courtyard as we watched guards dash about closing up camp and preparing vehicles for travel. The guards were smiling, something they rarely did, and I sensed something big had taken place to cause such a change in their demeanor.

Then, as before, we were loaded into the back of trucks for another move. But this trip was quite different from the one eight months earlier when we were herded into trucks, blindfolded, and our hands tied. Now, although we were roped together, we were not blindfolded, and instead of a thirty-six hour journey, the trip back down the mountain took only twelve hours. The driver drove like a madman as we bounced and jostled around hairpin curves and over rutted roads with no regard for our kidneys.

As we drove across a temporary pontoon bridge, to bypass the bombed out Paul Doumer Bridge on the Red River, I realized we were headed back to Hanoi.

CHAPTER 14

Goodbye Hanoi

AFTER YEARS OF WRANGLING to negotiate a peace settlement, and pressured by Congress and a war-weary American public, President Richard Nixon ordered an airstrike offensive known as the Linebacker II Operation in December 1972.

During the offensive, also called the "Christmas Bombings," twenty thousand tons of bombs were dropped on military targets in the area between Hanoi and Hai Phong. The bombings, the most precise the world had ever seen, led directly to the signing of the Paris Peace Accord on January 27, 1973, and brought about the release of 591 Americans held captive in North and South Vietnam.

I was returned to the infamous Hanoi Hilton and placed in Camp Unity once again with approximately thirty-five men, many of whom I had never seen before. Men who had been at Hoa Lo during our stay at Dogpatch told about watching B-52s fly over the city for three straight nights, dropping bombs while SAMs streaked through the night, lighting up the sky. It must have been one scary sight not knowing if a bomb would hit the prison. The prisoners said that on the first night guards ran around yelling and screaming. The second night fewer guards were seen, and by the third night, they saw no guards at all.

Now, the guards were back on duty. The prison showed no evidence of damage done by the bombings, and little had changed. The meals were still terrible and we were allowed out of our cells for only fifteen minutes to bathe and empty our latrine buckets. But there was a definite difference in the atmosphere, which hung heavy with a sense of urgency. Guards went about their duties at a hurried pace, scurrying about the camp carrying tools and buckets of paint.

The next day we were instructed to put on our long striped prison clothes and directed outside. Entering the courtyard, I saw hundreds of POWs. I couldn't believe my eyes. I knew there were a lot of us being held captive, but never this many.

We stood in what would have passed for a decent military formation as Rabbit read from a paper telling us the war was over, and we were going home. He explained we were to be released in four equal groups, and in what order we were to leave camp. The release of all the prisoners was to be completed by early March with prisoners in the South being freed during the same time frame.

Cameras were there, of course, to record the event, but the Vietnamese got a dose of disappointment when we didn't jump up and down, ripping off our shirts, overwhelmed with joy. After years of their lies we couldn't believe anything they said now. No one cheered. No one talked. We stood emotionless. Finally, a senior ranking officer gave the order dismissing us and we quietly returned to our cells.

I sat on my bed, wanting to believe my release was imminent, but could not allow hope to overshadow reality; I was still behind prison walls. The announcement, one we had waited for so long, just didn't sink in. Only when I was on a plane headed home would I believe I was really free.

Days passed without further harangue or pronouncement from the Vietnamese. Then on February 12, I watched through our tiny barred windows as the first group of fellow prisoners, dressed in tan jackets, dark pants and blue shirts, carrying a black tote bag, lined up, and walked out of the prison.

Seeing men leave in normal American clothes finally gave me hope, yet a voice inside my head warned me that this could be a ruse, a ploy to garner information from us. Holding to a glimmer of optimism, I wished them well and prayed for their safe journey home.

One of the Plums stated that if and when we were ever released we would go home in the order of our shoot down date. The only exception would be the very sick and wounded and the men shot down and injured during the Linebacker II Operation that previous December. I was to be in the second group to be released.

On February 14 a small group of us were taken outside to bathe. I noticed several men dressed in civilian clothes and some in foreign uniforms walking around the compound smiling and talking among themselves, but they never spoke to any of us. After we finished bathing, the guards took twenty of us to a different cell in New Guy Village adjacent to the courtyard. The cell had been cleaned and freshly painted, and the window bars had been removed. A couple of the visitors followed us into the cell and seemed to be inspecting the lodgings. This was very strange indeed.

A few minutes later Rabbit came into the room and announced that we were next being released. We were given new clothes, same as the ones I saw the first group of men wearing, and told to put them on. We were going home today.

Wait a minute. Something was fishy. There were only twenty men in the room.

James Pirie, who was the senior ranking officer among us, reminded Rabbit that we were supposed to go home in four equal groups of one hundred men.

"Where are the other men?' Pirie asked.

Ignoring the question, Rabbit said, "Put on clothes. Airplane come."

"No, this doesn't sound right. We aren't leaving out of the agreed upon order."

"That's okay. No need to put on clothes. Airplane broken, not come today."

The next day, another official told us once again to put on our clothes. "You go home today," he said.

With a stiff reserve, Pirie told him we didn't believe that what we were being told was legitimate and we refused to put on the clothes.

The camp official then asked us individually to put on our clothes. We all refused.

This continued for five days with the same answer, "We will not put on the clothes."

Pirie requested to speak to Norm Gaddis, now the camp SRO, since SROs Stockdale and Risner had left in the first group.

Pirie explained to Gaddis that we wouldn't go and told him why. Although the twenty of us were next to leave, we weren't leaving without the rest of the men in our group.

Gaddis then spoke to Air Force Brigadier General James C. Roan, who was there to ensure that the release went smoothly, and explained the situation. Roan told him he didn't know exactly what was going on, but did know the release was genuine. Gaddis then relayed to Pirie what General Roan had said.

Pirie stood his ground and restated our position to the camp commander; the men will not go, and the wrangling continued. Finally, Gaddis came to our room and asked if we would put on the clothes and prepare to leave.

Every one of us said, "No."

"I'm giving you a direct verbal order. Put on those clothes. You will go home today," he said.

We were suddenly faced with a dilemma—torn between wanting to go home and sticking to our promise to stay put until we left with our entire group. After all these years holding to our belief that we would not leave except in the order of our shoot down date, we were now being ordered to do otherwise.

Still reluctant, but faced with a direct order, we put on the clothes as instructed. Pirie told the Vietnamese we would march out in military formation, but if we saw even one camera, we would turn around and go back to our cell.

On February 18, 1973, I took one last look around the room wondering if I should take anything with me. Did I even want any reminders of this place? After some thought, I decided to take my tin cup, a compete set of long, striped clothes on which I had stitched my wife's picture, two packs of cigarettes, my bamboo fan, the German book written during my German class, and the house floor plans drawn under the tutelage of Danny Glenn. I stuffed them into my tote bag and zipped it closed.[1]

I should have been elated that I was going home, but in the back of my mind I half expected, and feared, the deal would suddenly collapse. When the door finally opened that last time, I walked out the same gate I had entered five years, eight months, and four days earlier, and never looked back.

Waiting outside was a faded green bus belching idling exhaust. With no cameras to record our departure, we boarded and were driven to the Gia Lam Airport about three miles from Hanoi. No one talked during the trip; perhaps we were still in a state of disbelief and shock.

At the airport, a Vietnamese official told us that our plane had been delayed. But we had seen the C-141 Starlifter, with a big red cross on the tail, circle and land. Now what were they trying to pull?

The driver continued down a side road to a building that appeared to be the headquarters of a Vietnamese fighter squadron. We filed off the bus in military formation and marched into a briefing room and filled two rows of seats.

A buffet table laden with food and drinks sat on a stage at the front of the room. A smiling Vietnamese official waved his hand toward the food and said, "Here's food. Help yourself. Need nourishment for long fight."

No one talked. No one moved. We saw this as a last ditch effort to take photos they could use to show the world their humane and lenient treatment; one last propaganda stunt; a fond farewell.

Every time they spoke to us we deferred to Pirie for his guidance. I sat watching the Vietnamese become impatient and frustrated at our lack of gratitude for their "kindness." Knowing there had to be cameras somewhere, we weren't about to be photographed accepting their hospitality. One of the Plums was "don't kiss them goodbye" and we weren't about to give them the satisfaction of showing appreciation for anything they did.

That was our last SIUTGA.

Finally, they gave up and said, "Your plane come now."

A less-than-cheerful guard conducted us back to the bus and drove us to the flight line where the beautiful C-141—our flight to freedom—sat waiting.

Stepping off the bus, I noticed a large crowd of Vietnamese civilians standing along the sidelines watching the pageant play out before them. That big beautiful bird flying into their capitol must have made a magnificent impression. I may never know what was going on in their minds, but by their expressions they seemed amazed to see American Air Pirates welcomed enthusiastically by U.S. military representatives. Supposedly we were defeated, blackened criminals, but there we were, being released in a grand and glorious

1. Memorabilia such as the plates mentioned in chapter 7, pictures of the POW's uniforms, General Mechenbier's drawings of the house he built in his mind, and the striped prison uniform on which he stitched the picture of Jerri, along with the shorts with the little red heart on it that his wife sent him, can be seen at the National Museum of the United States Air Force in Dayton, Ohio.

fashion.

Photographers and reporters were everywhere lining the tarmac. I recognized Frenchy, Bug, Soft Soap Fairy, and the Rabbit sitting behind a long table looking as though they had smelled something foul. I made eye contact with Soft Soap Fairy. It was hard not to laugh at his lemon-sucking expression as he sat there doomed to live out the rest of his pitiful life in this armpit of a backward country while we were about to fly home to air-conditioned freedom.

Representatives from the International Control Commission, made up of members from Canada, Poland, and India were there to oversee that the release and transfer went according to procedure. I learned later they were the same men I saw walking around the prison courtyard, and who followed us to our cell to inspect our living conditions. Too bad they weren't there years earlier. They would have gotten an eyeful.

We stood in line and waited as a Vietnamese representative called each of our names. There was no fanfare, no music, no flag waving, just a heaviness that hung over the flight line as everyone waited in anticipation.

"Mechenbier."

I stepped forward.

"Are you Edward J. Mechenbier?" the representative asked.

"Yes," I answered.

Satisfied that I was Mechenbier, he handed me over to the American colonel in charge of the delegation accepting us. I saluted and shook his hand, and then an escort took my tote bag as if I were too weak to carry it myself and guided me toward the airplane. It wasn't until I turned my back on the Vietnamese and started toward the airplane that I dared to smile.

Even though I was now in the hands of Americans, with each step I took toward that C-141, I still feared something might go wrong and kept praying, please don't screw this up guys. Just get us on the plane; get us on that beautiful airplane.

I climbed the ramp in a daze. Someone asked my name again, and then directed me to a seat. I sat waiting for lift off, afraid I might be dreaming and wake up to find myself back in my tiny, filthy cell with a rat staring at me.

There was a surreal silence inside the airplane as it taxied and then began rolling down the runway. Was this actually happening? Was this really the end of interrogations, bad food, pain, and degradation? Was this the end of waiting in fear, hearing the rattle of keys?

We broke ground from Gia Lam Airport in Hanoi, North Vietnam, at 1603 hours, but it wasn't until we were out of Vietnamese air space and over water and I heard the pilot say, "Feet wet. Welcome home," that I knew I was truly free.

Men went wild, whooping and shouting, clapping, and stomping. The cheers were deafening. Had anyone tried to squelch the noise, it would have been like trying to hold back the ocean with a fishing net. Some men laughed. Some cried. Some men danced in the aisle. If adrenaline could have been bottled and used as fuel, we would have had enough to fly us all the way home.

After the men settled down, I noticed the man sitting next to me wearing a black suit and sporting a crew cut. I stuck out my hand and said, "Hi, I'm Ed Mechenbier."

He turned to me and said, "Yeah, I know Ed, I'm Dr. Roger Shields. What in the hell have you guys been doing for five days?"

He looked annoyed.

"What do you mean?" I asked.

"You guys just about committed the biggest political blunder of the war."

I still didn't understand.

"You almost screwed up the whole deal."

I told him that we didn't know what was going on and we weren't going to leave until we had assurance that the release was legitimate.

I don't think my answer satisfied him. He got up and went to the back of the airplane.

The rest of the flight was filled with lively chatter and behavior unbecoming an officer. Some men goggled over copies of *Playboy* while others smoked real American cigarettes. Some savored the delicious flavor of coffee and tea for the first time in years.

Everyone talked at the same time wanting to know what was happening back in the good ole United States. We were told that women's fashion of mini skirts had come and gone. Darn! I would love to have seen that. We heard about the sexual revolution and that women no longer wore bras.

Now, you're talking.

One of the nurses overhearing our conversation interjected that Air Force women still wore bras, and with that took my hand and put it to her breast and said, "But it still feels the same doesn't it?" I was shocked. Just what had happened while I was away?

One by one we were taken to the back of the airplane for a cursory medical examination. I was told to lie down on a stretcher while a medic assessed my vital signs and asked me a series of questions. Did I have any broken bones? Did I have an infection? Blurry vision? Fever?

Noting that at one time or the other I had suffered from all of these conditions, the reality of my medical condition hit me. Just how sick was I?

Secretary of State Dr. Henry Kissinger, in a show of goodwill, broke protocol and requested the release of twenty men who became known as "Kissinger's 20." The Vietnamese gave him a list of prisoner's names from which he randomly chose twenty, unaware of the desire of the American prisoners to leave only with their entire group and in order of their shoot down date.

Dr. Roger Shields, from the Office of the Assistant Secretary of Defense for International Security Affairs, was in charge of planning and coordinating the repatriation operations.

The twenty men released in that group were Major Joseph S. Abbott, Navy Lieutenant James W. Bailey, Major James R. Berger, Captain John W. Clark, Captain Joseph Crecca Jr., Captain John O. Davies, Major Hubert K. Flesher, Captain Henry P. Fowler Jr., Major Donald L. Heiliger, Major Jay R. Jensen, Captain Michael C. Lane, Captain Kevin J. McManus, Captain Edward J. Mechenbier, Captain Joseph E. Milligan, Captain John H. Nasmyth Jr., Navy Commander James G. Pirie, Navy Lieutenant Joseph C. Plumb Jr., Navy Lieutenant Commander Frederick Purrington, Captain Herbert B. Ringsdorf, and Captain James R. Shively.

CHAPTER 15

Hello Freedom

TWO AND A HALF HOURS AFTER LEAVING VIETNAM, the airplane landed at Clark Air Force Base in the Philippines. Commander Pirie was first to disembark. The rest of us followed in alphabetical order, one at a time, as our names were announced, but Kevin and I stepped off the airplane together. We hadn't planned it that way; it just happened. I guess, after being joined at the hip, so to speak, for so long we had begun to think alike.

The first thing I noticed was how fresh and clean the air smelled. Even the sky seemed bluer. Kevin and I stood together and saluted the senior officers in the reception line, shook their hands and then turned to wave to the cheering crowd holding banners reading, Welcome Home and We Love You. What a warm feeling to see all those smiling faces.

It was hard not to rush over and say thank you, but we had been instructed to stay our distance. Perhaps the medical officials weren't certain what contagions we might transmit or maybe it was because we still carried the stench of Vietnam. Whatever the reason, we were hustled to a waiting bus and taken to the medical center where another sea of banner-waving well-wishers greeted us.

Doctors, Psychiatrists, and Dentists

My first stop was an examination room for a preliminary medical assessment to determine if I had any urgent medical needs. Finding nothing that demanded immediate attention, I was shown to my room and given a hospital gown and a striped, two-toned blue hospital robe, and then directed to a bathroom. I stood under the hot water scrubbing with sweet smelling soap and shampoo, washing away layers of Vietnamese filth. The clean, hot water felt heavenly and I wanted to stand there forever and let it rain over my tired, aching body.

Afterward, I was directed to the cafeteria. My mouth watered at the thought of real, hot American food, but because of my broken teeth, I was unable to eat solid food.

When the dietitian asked what I wanted. I told her I wanted a steak. I wanted a baked potato. I wanted corn, but couldn't chew.

"We can fix that. You can have exactly what you want," she said and smiled.

Apparently from the cursory medical evaluation on the airplane, the medics had reported my dental problems to the dietitian. While everyone else was having lobster, southern fried chicken, and pork chops, I got steak, potatoes and vegetables pureed in a blender to the consistency of baby food. You would think they might have blended the steak, potatoes, and veggies separately and presented them on the plate in three separate portions, but no, everything was blended together and served in a glass like a milkshake. I didn't care—I drank it through a soda straw, savoring the flavor.

Following that first meal I was immediately subjected to a battery of tests, both physical and mental. Doctors wasted no time poking and probing me from head to foot, taking blood and x-rays. They verified that my six-foot-two height had shrunk two inches due to the spinal compression caused during the ejection from my airplane, and I had dropped sixty-six pounds. Blood tests confirmed that I had contracted Hepatitis A, B, and C and would never be allowed to give blood; not surprising after my dip in the cesspool.

They also discovered I had four different types of parasites crawling around in my blood and intestines.

After the physical came a mental evaluation. Psychiatrists with a row of initials behind their names couldn't wait to get their hands on me. I could almost see them wringing their hands and chuckling like mad scientists with an evil gleam in their eyes as they worked their way through the interview, looking for signs to confirm their suppositions. Apparently they had conducted a study and determined that we would return home as basket cases. They told our families to be prepared for us to be suicidal and latent homosexuals; we would be indecisive, unable to make clear decisions for the rest of our lives. Most of us would be dead within five years because of the diseases we had contracted. Finally, we'd lose our hair.

Wow! What a future I faced. I began asking myself if I was as mentally and physically screwed up as they claimed. But they were the experts and who was I to question their findings?

As if I didn't have enough to worry about, they told me that oh, by the way, while you were gone for six years all your contemporaries have attended advanced schools, gained more experience and earned promotions. Some have made the rank of major one and two years below the zone. You'll never catch up. Now, what are you going to do with the rest of your military career, marginalized as it is?

Welcome home!

I suppose they were trying to help, but what they were really doing was planting doubt in my mind about my future in the Air Force.

With uncertainties beginning to surface, I told myself I would stay in the Air Force for two years, get the feel of things, and then see what happened.

Next, came a visit to the dentist. I'm not overly fond of dentists to begin with and recalling my episode with the Vietnamese dentist, I was not excited about repeating the experience. I nervously sat down in the chair with a case of collywobbles.

Although having my tooth pulled in such a primitive manner while in prison was excruciating, there was one saving grace to their method. They didn't try to anesthetize me. That would have put me over the edge. From the time I was quite young I have had a deathly fear of being given an anesthesia. In seventh grade I had two teeth pulled. When the doctor put me under the gas, I had a horrible nightmare that I was inside a dark sphere filled with water, rotating around and around on the end of an arm. The only sound was a drip, drip, drip. I was certain that if I were ever knocked out with gas again, I'd have that same nightmare.

I must have spent twenty-four of the seventy-two hours we were in the Philippines sitting in the dentist chair as the dentist pumped my gums full of Novocain and then drilled and filled, performed root canals, and capped my broken teeth. I was there so long I fell asleep in the chair.

After all the physical and mental probing came the inquisitors with their lengthy list of questions. Just as the Vietnamese had their blue book, I now faced Americans with a green one.

The first question they asked was if I was aware of other men still being held in captivity. They kept close watch as they documented my answers. They wanted information while it was still fresh in my mind and before it got filtered and cross-pollinated with other POW's stories.

Once on a roll they began bombarding me with questions about what it was like being held in a North Vietnam prison and about the treatment I had received at the hands of the Vietnamese. From the direction their inquiry seemed headed, I suspected they wanted to hear the gory details of what prison life was like and how it felt to be tortured. If spending five years and eight months in hell wasn't enough, their questions became another form of torture, and I wasn't going to humor anyone's voyeuristic appetite.

I know they were just doing their job, but asking me to relive those years was more than I should have had to endure. So, as I had done during interrogations while in prison, I began redirecting the session without fully answering their questions. I told them that the Vietnamese felt if they controlled our bodies, they controlled our minds. Then I talked about the clothes we were given and the environment in the different camps.

Next, came the safety officer who wanted a report six years after the fact about how the "accident" happened. What were the weather conditions? What was the ceiling at the time of the accident? What were the circumstances under which I lost the airplane? Was it mechanical failure? Were there things taught during flight school that would have mitigated the circumstances of accident?

Give me a break! Being shot down was no accident, but there was no place in his book to note that fact.

It was bad enough the dentist and psychiatrists had drilled me, now these guys were hounding me with inane and tedious questions to fill spaces in their book. The document ended up being seventeen hundred pages.

The Purple Popsicle

When I was finally allowed to return to my room, I found a note on my nightstand reading: "Sir, if you'll come to the radiation lab, I want to give you a purple Popsicle."

Grape was one of my favorite flavors and I hadn't had a Popsicle since I was a kid. Having been out of touch for so long, I wasn't familiar with the latest slang, and remembering the nurse's comment and action on the airplane, it struck me that perhaps a purple Popsicle might have another, more subtle, connotation. Perhaps the woman was making a seductive offer. I wasn't about to follow through, no matter how hot she was, but being the gentleman that I am, and not wanting to offend her, I put on my robe and slippers and padded down the hall to the elevator and found my way to the radiation lab.

When I got there, a lovely young lady, wearing a lab coat, was waiting for me holding a purple Popsicle. Whew! I dodged what could have been an awkward situation and got exactly what I wanted—a smile and a Popsicle. I found out after returning to Dayton, the woman was from Ohio and had hoped to use me to hook a space-available flight back home.

Later that night, arrangements were made for us to call home. The last time I had heard my wife's voice on the telephone was before I left Hamilton Air Force Base for Vietnam. Once I got to Vietnam there was no way to call home, so Jerri and I exchanged voice cassettes by mail.

The first question Jerri asked was about the circumstances of how I got shot down. She told me about a dream she had almost six years earlier, and described my ejection from the airplane exactly as it had happened. Her prophetic dream was unusual to say the least.

I assured her that I was okay, and she reassured me she was well, the family was fine, and I had a place to come home to. I was happy to hear that because some of the men had received letters telling them their wives had left them and had remarried. Guess the Vietnamese were right about one thing after all.

It came as no surprise that Kevin and I ended up as room buddies again during our three-day hospital stay in the Philippines. He was undergoing the same intense examinations and scrutiny, and the only time I saw him was in the evening when we returned to our room.

Worn out and exhausted after being placed under a microscope all day, I looked at the bed with its white sheets and fluffy pillow. I touched the softness of the sheets and sat down on a bed that actually moved under me. Once I laid down, it was as though someone had flipped a light switch. I passed out and slept until someone woke me the next morning some ten hours later.

These Shoes Don't Fit

Pressure had been placed on the Public Affairs Office to allow the returning POWs to talk to the media. But fearing someone might inadvertently say something that would jeopardize the safe return of the remaining prisoners, the Air Force was reluctant to allow any of us to give interviews before all the POWs had been released. So, I was surprised when an Air Force official came into our room and told me to hurry and get dressed; I was to give an interview to CBS, and the reporters were waiting down the hall. Shadows of Vietnam washed over me as I flashed back to guards telling me to get dressed; I was to give an interview for the troops in the South.

The officer handed me a new uniform, complete with shiny new captain's bars. Ordinarily, when an officer is promoted there's a ceremony with the new rank pinned on each shoulder by the commander, but I was handed my new rank on a hanger. That was the sum total of the ceremony—six years late.

I quickly donned the uniform and grabbed my new pair of shoes and put them on. They were tight, but I squeezed my feet into them not thinking much of it. After wearing only rubber sandals for years, I assumed my size eleven feet had changed shape and needed time to adjust to American shoes again.

A conference room had been set up like a small television studio with bright lights and several cameras on dollies. When the little red light flashed on, a reporter began the interview.

I had been cautioned not to say anything negative about the Vietnamese, the weather, the prison, the guards, the treatment I received, or the food, nor was I allowed to mention the names of anyone who was still being held. With these restrictions, all I was left to talk about was how happy all of us were to be free and how much we were looking forward to going home. A military officer was there to clarify any statement I made in case I misspoke. "What the captain means is . . . "

During the interview, I sat in an easy chair fidgeting, curling my toes, and wishing I could unlace the shoes to relieve the pinching because my feet hurt like the dickens. I got through the interview and went back to my room and told Kevin that I was going to have a hard time getting used to wearing shoes again. When I took them off, however, I discovered that in my rush to dress I had put on Kevin's size nines instead of my size elevens.

The second evening, we were given a few dollars and taken to the Base Exchange where employees had volunteered to keep the BX open after normal operating hours. I stood there in my hospital garb, surveying the aisles, like a kid in a candy store. I wanted everything, but didn't know what I wanted. I needed everything, but didn't know what I needed. Finally, I bought a Seiko watch to replace the one taken by the Vietnamese when I was shot down, and a piece of jewelry for Jerri.

My trip back to the States continued the next day with a quick refueling stop in Hawaii where, again, a large, cheering crowd greeted us. As before in the Philippines, Kevin and I got off the airplane together, stood together, and saluted together as cameras recorded the event.

Having determined in the Philippines that we weren't carrying any deadly diseases, we were allowed to approach the people. I walked over to the line of folks cordoned behind a rope and shook a few hands and thanked everyone for their support. I looked over at Kevin. He was hugging and kissing all the women, having a great time.

Then it was on to Travis Air Force Base in California. As I got off the airplane I saw two women holding a big sign that read, "Welcome Home, Ed Mechenbier." I walked over and shook their hands. As I did, one woman grabbed me and gave me a big kiss and said, "Say hi to Jerri." A little self-conscious, I wondered what the dickens was going on. All I could say was, "Thank you very much." I found out later that she had worked with my wife back in Dayton.

I said goodbye to Kevin at the airport with a promise that we'd see each other again soon and told him to take care of his ass. Knowing that Kevin's physical problems were worse than mine and that he didn't heal as fast, I had a foreboding that he wasn't going to do well over time. At the same time I was happy he was going home to be with his wife whom he had married only three months before being shot down.

The next leg of the trip took me to Scott Air Force Base in Illinois. There I transferred to a C-9 Nightingale medical evacuation airplane along with Charlie Plumb and Don Heiliger, two fellow POWs.

With fewer men on the airplane, there was less talk and more time for reflection. At each stopover—Philippines, Hawaii, and then Travis—I began to feel as though I was on a roller coaster. First came the joy of seeing banner waving crowds and bands welcoming us home. Then came the slow down as I got back on the airplane and was left to my private thoughts. Five years and eight months ago I was a young idealistic fighter pilot, full of hopes and dreams. Now, I wasn't sure who I was.

When I was first shot down and taken prisoner, my fears were that I might be killed by a bomb hitting the prison camp or beaten to death during an interrogation. I even feared dying from the bite of a poisonous snake. When nothing happened, the prospect that I might have contracted some fatal disease weighed heavily on my mind. Now, sitting on the airplane, I worried that I might be dead in three weeks because of some heinous malignancy. I knew I had medical problems and the full prognosis gnawed at me. Would all the predictions the doctors and psychiatrist made come true? All the worries I had buried for years suddenly came rushing at me, and for the first time I realized that I was scared.

Deciding ignorance was bliss, I pushed those thoughts from my head and began to worry about fitting back into society. What were things like at home? Would everything be different? Had Jerri changed? My range of emotion ran the gamut as I wondered what my life would be like in the future, if I had one.

As we departed St. Louis the pilot asked if we had seen the Gateway Arch. None of us had, so he circled St. Louis until they found someone on the ground to turn on the lights so we could see the Arch at midnight. All the fuss of seeing the lights on the Gateway Arch meant very little to me; I wasn't there when the wagons rolled westward. I didn't care about the lights. I just wanted to get home.

An hour later, my stomach tightened as the pilot announced we would soon be landing at Wright-Patterson Air Force Base. I had grown accustomed to the daily routine of being a prisoner and although it was terrible at times, I had become comfortable wrapped securely within the confines of the prison walls, knowing the rules and what to expect from day to day. Now, as the airplane landed, I looked out into the frigid night, not knowing what awaited me.

Jerri's Story

My heart raced as I sat in the car staring out the frosted window waiting for Ed's airplane to arrive. His flight should have landed hours ago and the sudden fear that something awful had happened terrified me. Finally, I saw the airplane land and taxi toward us. I watched as it stopped and its steps lowered.

I met Ed in April 1959 while working in the cafeteria at Saint Elizabeth Hospital where my job was to clear tables and carry dishes to the kitchen. When I first saw Ed washing dishes I knew I had to meet him. I would leave leftover pastries in the pantry where I knew he would find them, in hopes of getting a chance to talk to him. The cafeteria women got off work a half hour before the men, so when it came time to clock out, I cooked up a plan to have something to do until the guys finished their work and went to their lockers.

I was tickled the first time Ed noticed me. When he offered me a ride, my heart flipped; he was going to take me home. I felt like a queen riding in his 1950 Pontiac convertible. But he didn't take me home; he took me to the downtown Dayton bus stop to catch the bus.

We were both juniors in high school at the time. Ed went to Chaminade High and I attended Saint Joseph Commercial. Come May, I decided to ask him to my junior prom. I was a nervous wreck sitting on the basement steps in my mother's house twisting my hair around my finger and looking at the telephone trying to work up enough courage to dial his number. After he answered, we spent a couple minutes chatting about nothing, before I finally asked him if he would go to my prom with me.

Nonchalantly, he said, "Well, okay."

Two weeks later, he asked me out on a date but made it clear that it was to pay me back for asking him to the prom, not because he liked me. He must have liked me a little because we went steady throughout our senior year. As he was leaving for the Air Force Academy we had a big fight; he said I wouldn't wait for him. I told him I would, but he insisted I wouldn't. Obviously, he was wrong.

In the spring of Ed's first class year at the Academy, which would have been his senior year at a regular college, we became engaged. During the ring dance he took me out to the veranda and proposed. Remembering his response to me asking him to the junior prom, I said, "Well, okay."

I moved to Colorado Springs and stayed in the nurse's dormitory with my aunt, Sister Antonietta, a nun at the Penrose Hospital. My parents were very strict about that. I worked in the hospital there until Ed graduated.

Our wedding was beautiful. I felt like a princess. My little sister Bobbi was my flower girl and Ed's little brother Chuck was the ring bearer.

His brother and my sister were born one day apart in the same hospital in Dayton and the babies were in the same nursery. At age fifteen, Ed and I were too young to be allowed in the hospital to see the babies, so we had to sneak up the back stairs, but never saw each other there. I believe it wasn't the right time for us to meet.

Our first home, as a married couple, was at Vance Air Force Base in Oklahoma where Ed was assigned for pilot training. All the young couples were the same—poor, learning to play bridge, and trying to get pregnant. Our big night out was the last weekend of the month when we splurged and spent twenty-five cents on the All You Can Eat Spaghetti Dinner with an awful wine at the officer's club. That was our party time, and I loved it.

Our next move was to Davis-Monthan Air Force Base in Arizona. Our house sat on the edge of the desert surrounded by cactus, rattlesnakes, and great big, long centipedes. While Ed learned to fly the F-4C, I learned to keep house and be a military wife.

Then came his assignment to England. I had to wait until Ed found quarters before I was allowed to join him. I was a little leery traveling by myself, but made the trip just fine. We lived on an estate that was built prior to our American Revolution. The rooms were small with low ceilings. I laughed when Ed conked his head on the doorframe, and then he would laugh when I did the same.

We were in England only four months when his squadron got orders for Vietnam. As we left the base for the airport, I remember looking around the bus and thinking that not everyone there would be coming back. And as it turned out very few men did come home.

I returned to Dayton and Ed went on to Vietnam. I settled into a small apartment and busied myself working and worrying. I clipped and saved news stories about the war and put a large map of Vietnam on the wall to locate the towns mentioned in the news. I found Da Nang where Ed was stationed, and Hanoi where he sometimes flew.

On June 14 after saying my prayers, I went to bed and fell asleep thinking about Ed. That night I had a dream, a nightmare actually, that I was flying an airplane that was on fire, spinning out of control. Then I was floating to the ground in a parachute. I said a Hail Mary. I had a handgun, which, for some reason, I threw away. Then I couldn't move my arms because they were stuck to my sides. I awoke frightened, feeling certain Ed was in trouble. I looked at the clock on my nightstand; it read 1:15 a.m. It was 2:15 p.m. in Vietnam.

The next morning I told my mother about the dream, and told her if anyone in a blue car came to the door not to open it. Later that day, I flew to Chicago for a week of fun with my aunt the nun. We checked into the Palmer House and spent the day sightseeing and shopping. That night, as we prepared for bed, I heard a knock on the door.

As soon as I saw two men in Air Force uniforms my heart sank. I knew why they were there. They handed me a letter that read: "It is with deep personal concern that I officially inform you that your husband, First Lieutenant Edward J. Mechenbier, has been missing in action in North Vietnam since 14 June 1967."

Trembling, I asked if they could tell me where Ed had been shot down, but they didn't have details. Their job was to find me and relay the information. They did know, however, that Ed was in a four-ship formation. When they said a four-ship, I knew Ed had been shot down over North Vietnam.

There was more in the letter, but I couldn't read it with the tears welling in my eyes. All I wanted to do was get back home. We packed up immediately and went to the airport where we waited until morning for the first flight back to Dayton.

Once home I kept thinking I'd hear something soon, but I waited and waited. It seemed an eternity before I received a letter that confirmed Ed had been shot down.

> He was the aircraft commander of an F-4C aircraft on an operational mission in a flight of four aircraft. Coming off the target the aircraft was hit by hostile ground fire. The aircraft radioed that it was on fire and had lost all controls. The aircraft started rolling and the crew ejected. Two good parachutes were seen by other members of the flight and two emergency beepers were heard. The flight of remaining aircraft left the area, refueled and returned to search. This search lasted 45 minutes, but no visual sightings of the crew were made and no further beeper signals were heard. No further search was possible because of the hostile area in which the incident occurred. It is possible that your husband could have been captured.

Then on August 5 my parents and I visited the Shrine of Our Lady of the Snows near Saint Louis to pray for Ed. When I got home a letter was waiting stating that reliable intelligence sources revealed that Ed had been seen on the ground and that his status had been upgraded from missing in action to captured.

He was a prisoner of war.

Months passed before a brown cardboard box arrived at my home. From the return address I knew it was Ed's belongings. I felt numb as I looked at the package, realizing that what it contained might be the only thing left of him. My hands trembled as I cut open the box and folded back the two flaps. I didn't want any of his possessions. I wanted Ed.

I picked up the white tee shirt lying on top and held it close to my heart thinking I might feel him; smell just the barest scent of him, but nothing remained on his clothes and I began to cry. At the bottom of the package was a small box. Inside was a black sapphire ring. It gave me comfort knowing that Ed, even though he was busy fighting a war, took time to think of me.

In March 1971 almost three years after he was taken prisoner, I got my first letter. The Vietnamese had censored it heavily, cutting out most of the words, leaving me holding a piece of paper that looked like Swiss cheese.

Although I wrote frequently, I received only twenty letters in return during the entire time he was in prison, and most made no sense. He wrote about people I didn't know and vacations we had never taken. I found out later that he was sending cryptic messages hidden within the letter of men being held with him.

I was told how to write my letters on a six-line form. Ed had to use the same form in writing to me. Thanks to Ross Perot, our government finally got on board and allowed us to send packages. I was instructed that the parcel could weigh no more than 6 and 1/2 pounds, so I got out a baby scale to make sure I used every ounce. With his health my main concern, I sent him medication, knowing not everything would get past the Vietnamese.

I felt like a spy wearing a trench coat in a cloak and dagger movie as I tried to come up with clever ways to disguise the items I included in the package. I sent playing cards and wrote messages in the squiggles of the face cards. I sent One-A-Day vitamins mixed in with Red Hots candy and a pair of boxer shorts with a red heart embroidery on the rear. I sent Pringles, jellybeans, candy, and bourbon balls at Christmas. I also sent Band-Aids, aspirin, a washcloth, and small cans of ham and tuna.

Because my letters did not go through normal postal channels, when I wanted to send anything to Ed, I had to go to the main post office in Dayton and hand it over to the postmaster. From there it was hand carried by antiwar activists such as the American Friends Service Committee.

While standing in line to mail my letters, I read (upside down) the return addresses on packages being sent to Vietnam. This is how I found the names of other wives whose husbands were also POWs. Those living on the base knew each other, but we who lived off base were on our own. From the addresses I gleaned from packages at the post office, I made a list and contacted the women. I ended up with a list of ten waiting wives and we finally met in 1970.

The government didn't want the numbers of missing men to be publicly known, and I was advised to stay away from the base so as not to be a reminder to those serving there what could happen to them. The only Air Force contact I had was from a man at the base who would periodically visit with instructions and provide mailing forms.

I was rather naive back then and kept comparing Ed's situation to that of *Hogan's Heroes*. I expected Ed to do something fantastically heroic like marching out of the jungle with men following and music playing in the background. I had a lot of silly notions then such as why couldn't a *Mission Impossible* team sail up the Red River and pull him out of prison? Or why couldn't I go over there, knock on the door, walk right in and tell them I want my husband back? They're all little and I'm big. They'd be afraid of me. I had crazy ideas.

On Valentine's Day 1973 I went to Mass on my lunch hour and returned to my office just in time to take a phone call telling me that Ed was being released within the week. I knew this couldn't be correct. The Peace Talks had been on going for a long time and I had heard they had finally come to an agreement. The Peace Talks stated that prisoners were supposed to be released in four equal groups. Ed should have been freed according to his shoot down date, putting him in the second group, which wasn't to be released for two more weeks. This early release meant he was extremely ill or had been wounded. My heart dropped. I quit my job at the insurance company and waited with my parents for further word. We waited and waited but heard nothing for five days when I finally got word that he was on his way home.

The night of February 18 he called from the Philippines. When I heard his voice, the first question I asked after finding out he was okay was what happened when he was shot down. The story he told me mirrored my dream. Even the part where I was tied up and couldn't move matched with him being wrapped in his parachute lines. Every detail was the same. To this day I don't know what to make of my dream, but I do know that Ed and I have a special relationship.

The day Ed was to arrive, two men drove me to the base. They wanted to know if I wanted to have a party? Who would I invite? What did I want to serve? I told them that if there had to be a party, they could schedule any kind they wanted, and serve whatever they wanted. I couldn't think about that. I only wanted to think about Ed.

I remember getting dressed and looking in the mirror, worrying what Ed would think of me. I had lost a lot of weight while he was gone. It had been almost seven years. Would he still love me? Would he be pleased with what I had done during those years? Would he even appreciate what my time had been like during his absence?

I watched as the ground crew rolled the steps up to the airplane and Ed appeared in the doorway. I had been instructed to stay in the car until the greeting ceremony was completed and Ed had finished his statement to the press, but as soon as he finished shaking the dignitaries hands, I ran into his arms.

He was home at last.

World's greatest future fighter pilot at age two.

Mom and baby Gail, Jim, Tom, Ed, Dad.
Front row: Mary Lou, Jan.

Cub Scout, Albuquerque years.

Second Classman, 1963.

Cadet Mechenbier
Graduation photo, 1964.

Ed and Jerri in 1963.

Graduation day, June 3, 1964.
Chief of Staff, USAF General Curtis E. LeMay, "Father of Strategic Air Command."

Our wedding day, June 27, 1964.

First Lieutenant Ed Mechenbier checking out the T-38 at Vance AFB, 1965.

Within minutes after being shot down and captured, June 14, 1967.

Footlocks in Hoa Lo Prison—Hanoi Hilton.

Typical prison door with judas door.

Hanoi Hilton, Camp Unity, room no. 4.

The Pool Hall at the Zoo. Ed's room was the second door from the right.

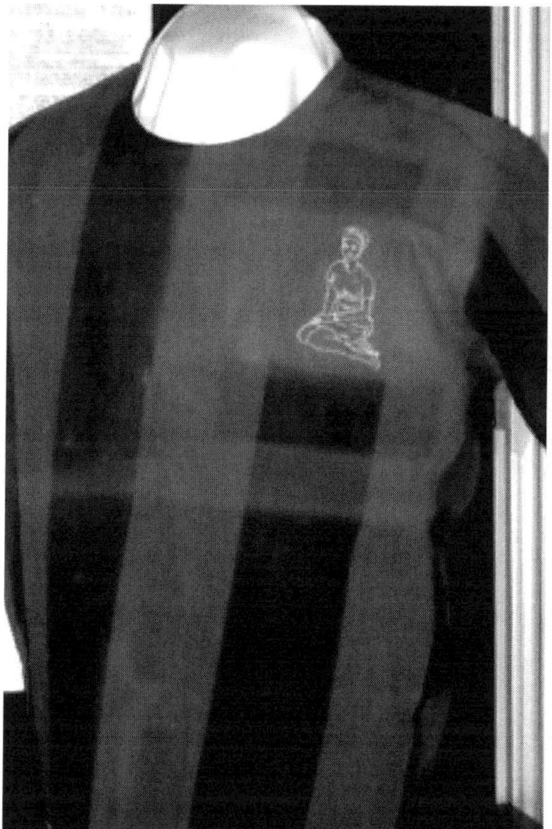

Prison shirt with my handy needlework of Jerri.

Typical prison uniform.

Photo on left: Finally a reason to smile. Day of release, February 18, 1973.
Photo above: Ed being escorted to C-141.

Joyous flight home aboard the C-141.

Ed and Kevin leaving the Philippines, February 1973.

Arriving home at Wright-Patterson AFB, February 21, 1973.

Big catch of the day in the Florida Keys, March 1973.

Throwing out the first pitch at Cincinnati Reds' opener, April 4, 1973.

Ed, President Nixon, and Jerri at the White House, 1973.

Crew Chief, Technical Sergeant Ray Kidd, checking out the Spirit of St. Louis II, 1978.

Captain Ron Brown following Freedom Flight #159, Randolph AFB, July 1979.

Ed and Robin Leach clowning around at the Dayton Country Club Dinner, 1999.

Lieutenant General Dick Reynolds, Ed, Unknown officer, Major Lori Foringer, and Colonel Mark Foringer performing Grog Bowl ceremony, 1999.

Ed, Robin Leach and Jerri at AFMC Commanders Wright Brothers Heritage Benefit, 2000.

Keeping watch during ceremony as the remains of American service members are placed in caskets at Noi Bai, Hanoi, 2004.

Repatriation Ceremony at Noi Bai Airport in Hanoi, May 25, 2004.

On left: Quiet reflections after repatriation flight on tarmac at Noi Bai Airport, Hanoi.

Below left: Caskets containing the remains of fallen service members inside C-141, Andersen Air Base, Guam.

Below right: Ed and wing commander laying wreath.

Steve Johnson, Tony Wilks, Rick Webster, Ed, Doug Jankovich, John Cherry, Scott Provost, and Mark Caudill in Guam.

Hanoi Taxi flight crew.

Here's to a terrific crew!

Traditional champagne wet down after fini-flight 2004.

Jerri, Mahli, and Kari getting into the action.

Bowing to my friend and mentor, Lieutenant General (Ret) George Rhodes.

The Mechenbier clan. Back row: Tom, Jim, Ed, Jan, Gail, Mary Lou. Front row: Chuck and Lora Jean.

Ed and Kevin in the cockpit of the Hanoi Taxi at Wright-Patterson AFB, May 2006.

F-4 flown in Vietnam, now in the National Museum of the United States Air Force. The two stars reflect Vietnamese aircraft shot down.

Whether flying high or low, they both go fast.

My Air Force family: Galen, Kari, Ed, Lindsey, and Bo.

Tai, Bo, Jerri, Ed, Mahli, and Kari before my retirement ceremony, June 30, 2004.

Furling my flag at my retirement ceremony. General Martin, Ed, Kari, and Bo.

Ed and Jerri at Ed's retirement party.

Ed and Kevin at Ed's retirement party.

Ed with USAFA Superintendent Lieutenant General John Regni (left) and classmate Jim Wheeler receiving Distinguished Graduate Award, April 4, 2006.

Ed by the judas door.

CHAPTER 16

Home at Last

WEDNESDAY, FEBRUARY 21, 1973, was the coldest, most bitter night of the year as media, family, friends, and service members waited on the flight line at Wright-Patterson Air Force Base for the airplane to arrive bringing home a man whom they had come to know as one of their own captured and held prisoner by the North Vietnamese.

It was 1:30 in the morning when the airplane taxied to base operations and stopped. I looked out the window and saw hundreds of people on the flight line shivering in the frigid night air. Scores of media with cameras and microphones waited against a backdrop of idling cars, spewing clouds of exhaust. I was humbled that so many people would stand out there in that deep freeze just to see me.

A public affairs officer had briefed me while on the airplane about what to expect when we landed. I was told that General Jack T. Catton, commander of Air Force Logistics Command and Lieutenant General Richard M. Hoban, his vice commander, would be on the ramp ready to welcome me home. My family would be waiting in an official staff car and would not be permitted to approach until I had finished a few remarks at the microphone.

When the door opened, I stepped off the airplane into a blast of Arctic air. The crowd cheered, their breath freezing in the night. Following a heartfelt welcome from the generals, I started toward the microphone, but before I could say anything Jerri came running into my arms. After that everything became a blur.

I vaguely recall walking over to the microphone and saying the first thing that came into my mind. "You're the greatest, every one of you. God Bless you all."

I remember being driven to the hospital in the back of a staff car with Jerri at my side. Once we arrived we were joined by my mom and dad and my dad's uncle, Brother Joe Fox, a Marianist Brother. We were then ushered to a consultation room where we were given a few minutes of privacy. I have no idea what we talked about. All I remembered was my Dad hugging me, my Mom wiping joyful tears from her face, and Jerri holding my hand, not letting go. After this short, supervised visit, everyone including Jerri, was shooed away under the guise that I needed to rest.

During those first four days in the hospital, doctors and psychiatrist took test after test as they continued to evaluate, dissect, and ponder my physical condition and mental state. Once they were satisfied that I wasn't going to throw myself off the nearest bridge and after they filled their notebooks with facts and theories, I was assigned to the hospital as an outpatient and allowed to join Jerri in her apartment overnight.

Although I was still somewhat in a daze, I did notice the rules of the medical profession had apparently changed because every nurse and every doctor I saw at the hospital was male. I don't know if they were afraid of what I might do, or whether they thought they might need brawn to control me. There was one exception—an older nurse that can only be described as the stereotypical battle-ax, someone you didn't want to mess with. She was the closest to a female I saw the entire time I was in the hospital.

When I finally was given time to sit down with my family and catch up on the last six years, I found out that my brother Tom had attended several colleges before finally graduating from The Ohio State University and embarking on a commercial real estate career. Jim had married and was practicing medicine in Columbus. Not knowing when I was coming home, my sister Mary Lou was in Florida planning her wedding. Jan had married and was living in Dayton. Gail had traveled from New Mexico to see me. Lora Jean was attending The Ohio State University and Chuck was still living at home. I am grateful no one asked about what I had been through. I guess they were waiting for me to offer information, which I never did.

Those first months were a whirlwind of activity. I had just left an environment where the daily pace was slow, with little variation. Now, I was suddenly thrown into a world where everyone was in charge of my life. Rather than being slowly reintroduced into this strange world, people rushed at me, pushing and pulling me in directions I didn't necessarily want to go. I was so torn between wanting to go forward and wanting to just go away that I don't recollect most of those first few days after getting off the airplane.

The Public Affairs Office, medical personnel, and my family made demands, telling me what to do and how to do it, leaving me little time to do what I wanted, at a pace I wanted. They all wanted something from me that I wasn't prepared to give, and I found myself agreeing even though I didn't want to.

I appreciated their interest and curiosity, but I wasn't thrilled at being expected to adjust to other people's agenda rather than being allowed to be me. I should have been elated to be home, but all I wanted was to be left alone and given time to adjust to a routine I could control. Instead, I was put on a treadmill, running at Mach speed.

One of the awkward situations I faced was finding out there were other men from the Dayton area shot down in Vietnam who didn't come home. Because the Vietnamese never released information about the men they held captive, the wives were given little or no information about the status of their husbands. They waited along with Jerri on that miserably cold night with one last hope that their husbands would walk off that airplane. Only one husband got off—me.

I felt guilty and awkward talking to them. They were hungry for information and wanted me to tell them what had happened to their husbands. Had I heard anything about them? Had I seen their husband in prison? Did I know what had happened to them?

What could I tell them? I didn't know if any of their husbands had even made it into a prison camp. I tried to build a picture they could carry for the rest of their lives, one that explained what it was like to fly in combat and be shot at. I spoke in generalities and softened a very hard and painful reality.

While being held in Camp Unity, Mark Ruhling and I had a long discussion about what we considered important in life. After much introspection and examination of conscience, we jokingly confessed our transgressions and decided we had our priorities screwed up and needed to overhaul our lives. We formed what we called the Weasel Club and agreed to abandon our selfish way of life and become better husbands, fathers, and men, and not indulge in frivolous pursuits.

Remembering my promise to the Weasel Club and in keeping with the old saying that "today is the first day of the rest of your life," I put the punctuation mark on my past and told myself that I was ready to begin a new and more productive existence. I reviewed and re-evaluated everything that I previously felt was paramount in my life and came to the conclusion most were no longer significant.

Having spent way too much time playing golf, I dug out my now dusty clubs Jerri had stored away, and threw them in the garbage can. I wasn't going to waste time playing golf again.

It was a noble vision.

Trips, Gifts, and Special Perks

I quickly learned that being a returning POW came with many perks. The United Voluntary Services gave me an engraved plate, the Ohio governor John J. Gilligan, gave me a pitcher made in Ohio that bears the State seal. I received a lifetime membership in a European Health Spa and a lifetime pass for two to any professional baseball game. When Ford Motor Company gave me the use of a car for a year, Beau Townsend at Stengers Ford in Dayton invited me to his dealership and I picked out a 1973 Mustang Mach II that cost $3,333. That was some set of wheels. A year later, Ford gave me the opportunity to buy the car, which I did for $1,800 and kept it until 1978 when I bought a Corvette.

One perk was a trip to Florida sponsored by the Key West Chamber of Commerce where Dick Ratzlaff and myself; along with Dick "Pop" Keirn, a B-17 Flying Fortress copilot, who not only was a POW in Vietnam, but also had the misfortune of having been shot down and taken prisoner during World War II; and our families were given the VIP treatment. The Key West community and Chamber of Commerce opened their doors to us with free meals and hotel rooms. The mayor gave us the key to the city and private citizens arranged picnics and events for the children.

Mayor Charles "Sonny" McCoy made us an honorary Conch, a term used to identify native-born Key Westers. One night we were drinking and watching the sunset from the rooftop patio of his home when the subject of golf came up. I said that I no longer played. That's when Jerri, in her infinite wisdom, plotted against me.

The next morning the sheriff came to our hotel, arrested me, and took me to Sears. He told me to pick out a golf shirt and shoes, and then took me to a local course for eighteen holes prearranged with the help of Mayor McCoy.

After finishing eighteen holes and enjoying a healthy time on the nineteenth watering hole, I was hooked. Playing golf was like absolution in a confessional—good for the soul, as Jimmy Buffett says in his song. That's when I realized that golf was a wholesome activity and not in violation of my promise to the Weasel Club. I love the game and can't think of a better way to be out doors enjoying life. Besides, the nineteenth hole holds many benefits. Yes, life was good once again, and I have been grateful for Jerri's insightfulness ever since. She is a wise woman.

The highlight of the trip was an excursion on a twenty-foot fishing boat in the Gulf for a day of deep-sea angling. The day of our outing the waves taunted us with what felt like fifteen-foot swells, but we weren't going to allow a little rolling water stop us—we were brave fighter pilots after all and besides, the media was watching, we had to show our mettle.

Never having been much for boating, it wasn't long before the heaving waves matched my heaving stomach and I became seasick. When I did get a strike, I begged the fish to spit the hook. There was no way I was going to reel in that sucker and not puke. All I wanted to do was put my feet on solid ground. Perhaps the ordeal wouldn't have been so bad if I had actually caught something praiseworthy, but all I had to show for my suffering was a fish no bigger than bait for a Barracuda, and not even good enough to be made into cat food.

A local newspaper reporter was waiting on our return to take a photograph of me proudly holding my big catch of the day. Luckily, the picture was in black and white. Otherwise the public would have seen the world's greatest fighter pilot in various shades of green.

But I wasn't finished with boating just yet. Bill Davis, a local citizen, invited Jerri and me to a day of sailing on his Invicta fiberglass sloop. Although I had never been on a sailboat before, remembering what Read Mecleary had taught me at Camp Unity, I felt well prepared to take the helm and sail her through the Keys. We navigated the Gulf for a couple hours as I demonstrated my knowledge of the binnacle, cringle, drogue, hawser, and jib. It was a great experience.

When it came time to return to the marina, I expected to relinquish control, but because Mecleary had taught me so well, I handled the boat like a professional sailor and Bill allowed me to dock the boat. I called down the sails to reduce our forward speed and began maneuvering the boat under wind power only. Bill got all excited and said, "This is going to be fantastic. You've aligned the boat perfectly. We're going to be able to put her into the slip without ever having to turn on the auxiliary power. Let me take over."

"Aye, aye, Skipper. You've got it," I said, and stood aside.

Bill took the helm and, not taking into consideration the depth of the water at low tide, ran the keel aground a minute later. We were stuck a hundred yards from the slip, and even using the auxiliary power, it looked as though we were going to be marooned there until the tide came in. I had relinquished the helm in the knick of time. Bill had a good path but the wrong time of day for the water level. Although, I had the boat set up perfectly, he screwed up my grand entrance into the harbor.

Despite the snafu, my seamanship must have impressed Bill because he invited me to help him pilot the boat to Baltimore. Sadly, I couldn't take him up on the offer because I was due back in Cincinnati to throw out the first pitch for the Reds' 1973 season opener.

Perhaps I should have helped him sail the boat north because I heard later that upon leaving Key West, for some unknown reason, Bill headed west and ended up wandering around the Gulf of Mexico for five days.

Having always been a fan of the Big Red Machine, I eagerly agreed to the invitation to throw out the opening pitch against the San Francisco Giants. Before the game, I was invited onto the field to meet Sparky Anderson, Johnny Bench, Joe Morgan, and Pete Rose. That was exciting. Then Jerri and I were escorted to an official box behind first base as reporters and cameramen crowded around to photograph me biting into my first hot dog and sipping a soda.

When it came time to throw out the pitch, I was introduced to the sell-out crowd as the stadium rocked with applause. I wasn't too thrilled about all the publicity but I did what I came to do. I hadn't lost my pitching arm from my high school days having played in the Dayton Amateur Baseball League, and heaving cinderblocks in prison had kept my arm in shape, so I didn't shame myself when I pitched the ball to Johnny Bench.

I was a bit embarrassed, however, when fans asked for my autograph. I couldn't believe anyone cared who Captain Ed Mechenbier was. They only knew I was a Vietnam POW and that was supposed to mean something, but I scribbled my name on whatever they handed me and thanked them for their support.

During the seventh inning a little boy approached and shyly handed me a piece of paper. "Sir, would you sign your autograph?" he asked. I signed and handed it back. He looked at my signature and then up at me and said, "Who are you any way?" I loved it. It was refreshing to see an honest face.

Despite all the whoop-de-do, I actually ended up enjoying the event even though the Reds lost.

The recognition continued as the POWs were made honorary citizens of Dallas, Texas, and given a ticker-tape parade with a grand marshal, colorful floats, marching bands, braided-tail horses, and clowns driving tiny cars—the whole shebang. As honored guests at the Cotton Bowl, we watched Bob Hope, with his familiar brand of humor, play to an audience of seventy thousand. Tony Orlando and Dawn also debuted their song, "Tie a Yellow Ribbon Around the Old Oak Tree" which became the anthem for people waiting for loved ones in the military to come home.

That summer an officer from the Office of Special Investigation requested my presence at Maxwell Air Force Base in Alabama for a week of indoctrination and updates. Once again I was questioned about events in Vietnam. They showed me photographs of camp commanders and guards from the various prisons, wanting me to identify them. Before long I began to realize the real reason behind the interview—they were trying to nail the Vietnamese for war crimes.

I began to detect a witch-hunt when they started asking if I was aware of any inappropriate activities by fellow POWs or if I knew of anyone who had breached regulations. Apparently someone had made allegations that not everyone had conducted himself according to the Code of Conduct, and the Office of Special Investigations was digging for information, trying to hang the offenders.

Dinner With the President

During that summer all the former POWs were invited to Washington for a formal dinner at the White House. That was when the president's residence was open without restrictions. Jerri and I were wandering around the family quarters when someone walked up behind us and said, "Would you like to see the Lincoln bedroom? We turned to see President Richard Nixon standing there dressed in tuxedo pants and shirt with no tie. He told us to look around, and then wandered away. Imagine our astonishment as I looked at Jerri and she looked at me, both of us wondering, was that who we thought it was?

An elegant dinner was held on the south lawn under a huge tent with china, crystal, and candles decorating each table. Unfortunately it had rained that afternoon making the ground quite soggy and the air humid. But the president was gracious and stood while photos were taken of us standing together. Jimmy Stewart was there and I got to kiss Joey Heatherton and Jerri rubbed Sammy Davis Jr.'s head.

This gala affair was the first time all the POWs were together and I finally got the opportunity to put names with the faces of men I only knew by their tapping on the walls. It was nice to meet Bob Shumaker face-to-face. I'm grateful to this day, despite the dangers, that he was determined to thread that tiny wire through the rat hole to get the tap code to me. His courage went above and beyond and his effort meant something very special to me.

To many, I suspect, dinner at the White House with the president of the United States was a big deal, but I felt as though I was a commodity being herded around and shown off to the masses like a prized heifer at a cattle show. I was grateful but all I wanted was time to myself to ease into my new world. I don't think anyone fully understood that we had just returned from hell and now they were hoisting us on their shoulders, hailing us as heroes. It was like coming out of major surgery and being asked to dance the watuzi. Our minds and bodies were still in shock unable to fully appreciate dinner with the president.

During our imprisonment the Voices In Vital America organization had started a bracelet campaign and after our release many of my fellow POWs began receiving bracelets with their rank, name, and date lost etched into them. Jerri hadn't wanted any part of the program and wouldn't allow my name to be put on a bracelet. She told me she wanted to protect us and didn't want all our information out there for the whole world to latch onto. I did, however, receive hundreds of letters from people saying they had prayed for me and were happy I was home. Unlike some of the guys, I didn't receive any marriage proposals or job offers, however. When I got time away from all the required public appearances, I answered every one of those letters and did have a few bracelets made for selected friends.

Our First Home

In August of that year Jerri and I bought our first home, a beautiful house in Enon, Ohio, near Springfield. As soon as we walked into the house, Jerri said this was *the* house, and she wanted it.

It was a great house with redwood paneling, two huge stone fireplaces, and a balcony off the bedrooms. It sat on a three-quarter acre lot with a copse of trees taller than the roof. We built a nine-hundred-square-foot family room that had a bank of thirteen-foot-tall windows. We lived there for seventeen years and raised our children there. The only reason we sold the house and moved to Beavercreek was for a better high school system for the kids.

In 1990, after coming home from a TDY assignment, Jerri told me she thought we ought to move so the kids could go back and forth to school more easily. At that time, they were riding some twenty-six miles on the school bus. The next day we went house hunting. Jerri announced that she wanted to drive, took the wheel, and headed directly for the place she had already picked out near Xenia. Apparently she had scoped out a house there and had made up her mind this was the one she wanted.

We put down five thousand dollars earnest money, but before the deal went through, Jerri found another more suitable abode in Beavercreek and invoked her right as a woman to change her mind. We lost the earnest money, but in the end it was worth it for the beautiful home we now have.

CHAPTER 17

Catching Up

IN 1968, MOVIEGOERS GOT THE BEJESUS scared out of them watching Mia Farrow in Rosemary's Baby. *Robert F. Kennedy was assassinated in Los Angeles, and Richard Nixon was elected president of the United States.*

In 1969, the New York Mets beat the Baltimore Orioles four out of five games to win the World Series and more than four hundred thousand concert-goers sat in mud, smoked pot, and danced at what became known as the greatest moment in popular music history—Woodstock.

In 1970, Ohio National Guardsmen fired into a crowd of students on the campus of Kent State University, killing four and wounding nine others. "War" was among the Top 20 songs of the decade, reflecting the sentiments of the nation.

In 1971, the last cigarette TV commercial was broadcast on the Tonight Show *hosted by Johnny Carson. China was admitted to the United Nations, and Joe Frazier beat Muhammad Ali in fifteen rounds.*

In 1972, the Washington Post *broke the story about the Watergate break-in at the Democratic Headquarters in Washington, D.C. The Dow Jones closed above 1000 for the first time in history, and Richard Nixon won reelection in a landslide victory with 60.7 percent of the vote.*

The changes I experienced after returning home caught me by surprise. I soon discovered, as Bob Dylan said in his song, "The Times They Are A-Changin.'" They certainly had. If you live through a slow evolutionary progress of social, cultural, political, and religious changes you don't necessarily notice the radical transformation. New inventions and technology become part of your life and you grow along with them. But if you are suddenly exposed to a Droid-X and all you've ever known is a rotary phone hanging on the wall, that revolutionary change can be overwhelming.

That's pretty much how I felt returning to a world that had moved on without me. While I was held prisoner in Vietnam, technology had gone from a slide rule to a multifunction calculator that spit out answers to complicated equations within seconds; audio cassettes had progressed to videocassettes; cash registers to barcode scanners; and the artificial heart was the new medical hope.

Social changes permeated every part of society. The hippie movement had brought about Flower Power, mystic religions, and free love. A new generation was wearing bell-bottom pants, love beads, and go-go boots. Men had long hair and women were breaking the "glass ceiling," while racial integration attempted to bring social justice to all.

Moral values had changed. The media allowed more cleavage on television, and language had become looser. The use of the "F" word on television was common, drugs and teen pregnancies ran rampant and women thought nothing of wearing skimpy clothing in public that left nothing to the imagination. The total disregard for manners and good taste was shocking.

Another change I noticed was the lack of regard for people in authority. The respect for the office of the president had eroded. When I left the states, people referred to the president as Mr. President or Mr. Johnson. When I returned, the press referred to the president as Tricky Dick. Doctors, lawyers, teachers, and members of the clergy who were once held in high esteem were pushed aside and replaced by rock stars, movie stars, and overpaid athletes.

In my youth, becoming a millionaire was earned through years of hard work. When I came home, millionaire status was achieved in one season by throwing a baseball ninety miles an hour or a football fifty yards, or by standing on stage and screaming to earsplitting music. I've never understood the adulation ascribed to them.

I noticed the Catholic Church had remodeled itself. In 1967 the Mass was recited in Latin and the priest stood with his back to the congregation, facing the altar, as if leading the service. No matter where you were in the world, you knew and understood what was going on during the liturgy. I remember attending a Catholic service in Bangkok and being able to follow along in Latin. The only vernacular was in the sermon and the announcements of wedding banns and Bingo on Friday night.

In 1973 Mass had become a social event with the congregation shaking hands and hugging. I couldn't tell if the service was Catholic, Episcopalian, or Baptist. The priest entered the church by strolling down the center aisle with an entourage of laymen. Hymns were called songs and the priest acted as though he was at a Saturday afternoon backyard cookout. Lay people helped serve Communion and gave the readings, and the priest was no longer called Father Kovach, but Bob. I can't buy into the changes made by Vatican II. To add insult to my faith, I had to look in the entertainment section of the newspaper to see where to go the church on Sunday.

Saying No

I was almost always agreeable to speaking requests and am reluctant to tell people no. But I did say "No" twice, however.

Jerri told me that while I was gone whenever there was news from Vietnam, WHIO Channel 7 in Dayton invited all the wives from the Dayton area whose husbands were listed as POW or MIA to the station to watch film footage of the war and offered them cookies and punch. The station would run the film many times, stopping it often so the women could scrutinize each frame for any sign of their husbands. The *Dayton Daily News,* the *Journal Herald*, and Channel 2 did absolutely nothing for the wives. They wouldn't even carry news about the war from their own affiliates.

As the POW/MIA movement began to gather steam, Jerri went to the *Dayton Daily News* to ask them to run a story, but they refused. She made the comment that if someone burned a flag they'd cover that. They agreed they would have.

I had been home only a few days when I began receiving invitations to speak to service organizations such as the Rotary Club, Kiwanis, and church groups who wanted to hear "my story." The Officers' Wives Club held a luncheon and invited four of us to be part of a panel to answer questions about our POW experience.

I was also asked to speak at the Engineer's Club, where the master of ceremony happened to be the news anchor for Channel 2. After wrapping up my speech, Mr. Slick came up to me and said, "Hey, we'd like to do an interview with you." After hearing how Channel 2 had treated the wives, it gave me great satisfaction to say no, and then I told him why.

A couple weeks later, Phil Donahue, who had worked at that same station and who was very much against the war, invited me to be a guest on his show. Again I said no. I wasn't about to go running into his arms, either.

I gave back to the people who took care of my family and I've had a close relationship with Channel 7 ever since. I have a great regard for the leadership at that station.

Back in the Cockpit

Soon after returning home, Major Joe Ashey, who went on to wear four stars, and was with the Air Force Personnel Center at Randolph Air Force Base in Texas at that time, asked me what I wanted to do with my career and suggested that I could have any assignment I wanted. Wow, that was great! I felt as though I had just won the lottery. My mind raced thinking about all the opportunities being afforded me. My dream of becoming a lawyer still lingered in the back of my mind, so I told him I wanted to go to law school and become an Air Force lawyer.

"We can't do that. Besides you're too old, at thirty-two, to go back to college. It would be easier to lower the subway fare in New York City than to find you a slot in law school," he said.

"But you just told me I could have any assignment I wanted," I said.

"What is your second choice?"

"In that case, I would like to go to the University of Miami, earn a masters degree in physical education, and go to the Air Force Academy as the golf coach."

Knowing that Academy instructors were considered the cream of the crop and being on the staff there was a good way to increase my chance for promotion, I considered the job a good fit. Besides being able to play golf everyday and be paid for it was a bonus.

"You're joking," he said and sat back, folding his arms across his chest.

With my first and second choices shot down, I asked what he wanted me to do. He said going to Davis-Monthan Air Force Base and becoming an A-7 Corsair instructor pilot was my best option. He told me that from there I could attend the Air Command and Staff College. He had just offered me a "non-starter" career path where I would be chasing after my contemporaries.

Accepting that assignment would have put me in a job with a combat squadron that was eligible for TDY 180 days a year. That didn't set well with me. After being away from home and my family for so many years, I didn't want to be in a unit that was gone for long stretches of time. My other choice was to shuck it all and leave the Air Force.

When Major Douglas Cairns, who was my host officer and also a 1961 Academy graduate, heard I was thinking about leaving the military, he offered me a job with the 4950th Test Wing at Wright-Patterson Air Force Base as a test pilot in the fighter branch. There should be an additional motto for Air Force Academy graduates that says, "We take care of our own."

Being a fighter pilot was great and, except for getting shot down, I was good at it, so I accepted Doug Cairns's offer and began a journey of legendary feats.

Along with all the negative predictions the Air Force psychiatrists made about our physical and mental condition, they also had erroneously preconceived ideas about what we would be like as pilots. Former POWs were not allowed to fly solo for fear we might be suicidal and fly the airplane into the ground.

In March of 1973, to make certain that didn't happen, the Charging Cheetahs of the 560th Flight Training Squadron at Randolph Air Force Base in Texas began the Freedom Flyer program for repatriated POWs. The primary purpose of the program was to reintroduce pilots to flying and ensure that their last flight in a military aircraft wasn't the one in which they were shot down and captured. The program sounded great, but after hearing horror stories about how former pilots were being watched for signs the psychiatrists had anticipated, I didn't want to go to Randolph and be subjected to the kind of scrutiny where I was gawked at as if I were a two-headed goose.

I wasn't prone to suicide and I certainly wasn't about to take someone down with me—I had had enough of that on June 14, 1967, thank you. I knew exactly what I wanted—to continue flying, and I didn't want to go to Randolph and put up with the craziness where I would be told, "do this," "do that." I'd had enough of that, too.

Since there was no regulation that said I had to go to Randolph, and given the opportunity to be back in the cockpit and fly with the 4950th, I jumped at Major Cairns's offer and made my reentry into flying in Dayton. Even if there had been a regulation requiring I had to go to Randolph, and I had said, "No, I ain't gonna go," what were they going to do if I didn't, put me in jail?

Before I could get back into the airplane, however, I needed a refresher course on the instruments and controls. My first time back at the stick was exciting. I had no flashbacks to my last flight when my burning airplane was corkscrewing through the sky. No nervous sweat; no reservations. I strolled out to the flight line and approached the airplane. Walking up to an F-4 in 1973 was like greeting an old friend and we were about to pick up where we had left off. It was as if the airplane had been waiting there for me like an old faithful dog for its master.

Sitting in the cockpit again felt like I had never been away. Nothing had changed. The smell of the cockpit was the same, the feel of the seat, the same, even the control panel looked as it had six years ago.

Major Jim Brenholdt, chief of the fighter branch who later became a lieutenant colonel, was my instructor pilot. He walked me through the preflight routine, marking off the sixty-some items in the spiral-bound book as I checked each instrument and switch against the list. The rule is that no matter how proficient you are as a pilot you always use the checklist. It would have been easy to look around the cockpit and make sure all the shiny switches were on and the rusty switches, that were seldom used, were off, but that wasn't the way I operated. Not following the checklist got a lot of pilots in trouble. There was no easy way around it. You just worked through the process.

With the checklist complete, I pushed the throttle forward and began rolling down the runway. My pulse raced as I felt the 17,900-pound thrust of the turbojets come alive beneath me. God, what a feeling!

The wheels lifted off the pavement and I was once again sitting behind the stick of a jet aircraft where I belonged. We flew to the restricted area near Peebles, Ohio, where the test wing conducted missions, and then returned to the traffic pattern at Wright-Patterson where we did a few practice approaches and landings. What a joyous moment! That's when I knew I still had it and was still the world's greatest fighter pilot.

The 4950th Fighter Branch had five pilots and eight aircraft; two T-39 Sabreliners, a T-37 Tweet, an F-4D Phantom, a hard wing F-4E, two RF-4Cs, and an F-100 Super Sabre. Working for the 4950th with so many aircraft available turned out to be a virtual flying club with a license to be a kid again. Obviously, we got a lot of flying time conducting tests, but we also got to do some truly exciting flying when there wasn't adult supervision.

The decision to join the 4950th put me back in the cockpit where I got eighty flight hours a month flying the F/RF-4 and the T-39 Sabreliner—a twin engine, six passenger, two crew executive airplane used for training and as a test bed for electronic pods.

My main job was to conduct tests on the various sensors and trackers on the T-39 side emergency escape hatch. With the turret mounted on the side, we flew behind other aircraft to collect their infrared signature while an engineer sat in the back monitoring the testing in real time. I also tested the F-4 variants used to carry experiments internally and in a pod, and conducted electronic and photo reconnaissance pod testing.

Knee Surgery

During captivity, calcium had built up around the pin that had been inserted in the leg I injured playing intramural football at the Air Force Academy.

Now, after eight years, my leg began bothering me and the doctors at Wright-Patterson decided the pin needed to come out and my leg fixed properly. Going through the paperwork at the hospital would alert the flight surgeon of my injury who would then take me off flying status for an undetermined time. I couldn't let that happen. Jim Greenfield an orthopedic surgeon, Marshall Brummer in internal medicine, and Julian Furiorni of general medicine, characters straight from the TV show *MASH*, bypassed the paperwork and snuck me into the medical center to perform the operation.

Nurses prepped me for surgery at 0 dark 30, the doctors removed the old pin, fixed my leg, stitched me up, and then discharged me that afternoon in a long-leg cylinder cast with orders to be careful. In almost no time, I was back flying with no record of ever undergoing surgery. Luckily, I was flying the T-39 most of the time, and unlike the F-4 where I had to swing my leg up and over the side of the airplane to get in, all I had to do in the T-39 was walk up a couple steps, open the door, and step in.

Golfing in the South

Since I had flown the F-4 for years, when the test wing began transitioning from the F-100 to the F-4, Brenholdt made me the F-4 instructor pilot, and in June I was sent to Shaw Air Force Base in South Carolina to finish my recurrency training.

Jerri had accompanied me on the trip to Shaw and, with my love of golf restored, and my supportive wife, who had a pretty good golf swing and score to match, we played every chance we got. We found a golf course on base not far from our quarters and set out for the first tee, pulling our clubs along behind us.

Light fluffy clouds dotted the clear blue sky and the weather forecast predicted only a ten percent chance of rain—a perfect day to hit the links. Somewhere between the third and forth hole I noticed the sky beginning to darken but thought little of it. On the fifth hole a light sprinkle began dotting the fairway. By the time we finished the sixth hole the sky had turned an ominous gray-green, and before we could putt the seventh green, it let loose a fierce downpour.

Normally, we rented a cart but that day we had decided to walk the course and work off the calories from the good southern food we had consumed during the previous week. Looking around, we realized we were too far from the clubhouse to walk back without getting soaked or stuck by lightning, so we took cover under the first refuge we found, a ten by ten open-sided shelter, to wait it out. Figuring the rain wouldn't last long, we stood huddled together there for three hours, watching the water rise around us. We almost drowned in that ten percent chance of rain.

With my training complete we headed back to Ohio when, just outside Columbia, South Carolina, we saw a sign announcing a golf course two miles ahead. Not one to pass up a chance to play, we decided to try our luck again. It was a sunny day with not a cloud in the sky, and the weather forecast promised no chance of rain. We said why not, and pulled into the golf course. When we asked what the green fees were to play, the golf pro attending the shop told us to pay him when we got back.

I told him we planned to play eighteen holes, but he insisted that we pay when we finished the game. This was highly unusual, but hey, he was the boss. Maybe they did business differently in South Carolina.

We teed off expecting a great time. The humidity was a little high but not unexpected for South Carolina in June so we started off on a day of enjoyable golf. The course was beautiful with long, straight fairways, well-kept, lush greens, and just enough doglegs to make the game interesting. The lack of people ahead and behind us promised an enjoyable and leisurely round.

We had made it to the third tee when I began to notice the temperature rising. I unbuttoned the top button on my shirt and addressed the ball. By the fifth hole, we had worked up a good sweat, and Jerri was showing signs of slowing down. Before we reached the tenth hole, Jerri and I had had enough of the South Carolina heat and no matter how alluring the course, we couldn't continue. Disappointed not to have finished, we dragged ourselves back to the clubhouse expecting to pay for eighteen holes.

There, we found the golf pro standing behind the counter laughing, "Now you know why I didn't take your money up-front."

With my additional training at Shaw complete, I now held the same title of research test pilot and flew the same programs as other test pilots, and also wore the same orange flight suits. But not having gone through the year-long USAF Test Pilot School at Edwards Air Force Base in California to earn the AFSC 2865 and the Golden Arm ranking, I was qualified for only the "Golden Arm" nickname with Reprobate Status, a tongue-in-cheek term of endearment and mocked scorn. Test pilots have always had a swagger earned by having completed the lengthy and rigorous curriculum, but my swagger had a slight blip in the step.

I worked with some terrific test pilots but one in particular became a good friend. Larry Bogemann, the world's second greatest fighter pilot, was a *real* test pilot and owned a TR3. I drove an MGB at the time. Always being up to a challenge, as winter approached, I dared Larry to see who would be the first weenie to put the top up on his car. We drove through rain, sheet, and snow each day to work bundled up like Eskimos as the weather became colder and colder, neither giving in. It was mid January before Larry finally conceded.

The next summer, Larry and I were on a local test mission each flying an F-4C. We had a photographer along that day and Larry decided he wanted a picture of his house from the air with an F-4 framed in the foreground. With me on his wing, Larry directed the photographer, who was in his back seat, when to take the photo. We made three passes and Larry would say, "Now," and the photographer would snap the camera. Unfortunately, Larry's directing ability was slightly off and the photographer missed his house each time. He ended up with great photos of the neighbor's houses.

Larry may have been a great test pilot and a wonderful guy but was lousy at directing a photo session.

A Special Christmas Gift

Just before Christmas my first year home I received a small UPS package from Boston. The name on the return label was familiar—Tim Sullivan, one of my fellow POWs. Inside, wrapped in white tissue paper, was a Newburyport pewter mug promised during our virtual Christmas gift exchange back in Camp Unity. The inscription on one side read, "North Vietnam 1970." On the other side was, "To the Archbishop who could take them or leave them." Tim had remembered my attitude toward smoking. I wasn't addicted to cigarettes and could take them or leave them.

It was a handsome mug about six inches tall with a curved handle set on a round pedestal base. I cherish that mug filled with so much meaning and display it in a place of honor in my office.

Likewise, I sent Galand Kramer, who was stationed at Eglin Air Force Base in Florida at the time, the 240 Baby Ruth candy bars I had promised. I included a note saying that I hoped the candy didn't melt on Christmas day.

CHAPTER 18

Test Pilot

NOW A FULL-FLEDGED TEST PILOT, I was deployed along with Timothy Westover, grandson of Oscar Westover for whom Westover Air Force Base in Massachusetts was named, as my copilot to Lambert Field in Saint Louis, which is co-located with the McDonnell-Douglas plant, to flight test the Tactical Early Warning System for the F-15 Eagle fighter aircraft. The day I arrived the weather was dog trash, turning the runway into a sheet of ice. The tower reported that braking action was minimum to nil but the runway had been sanded and we were cleared to land.

What the tower didn't realize, however, was the big commercial jets had blown off the sand during their takeoffs and landings leaving the runway with undetectable patches of ice. Without the antiskid or a thrust reverser system on the T-39, as I landed at a hundred miles an hour one wheel hit a dry patch. With lightning fast reflexes, I slammed on the brakes and locked that sucker up. After making a 360-degree donut in the middle of the runway, the airplane came to a stop. The tower, watching my airmanship, radioed to ask if we needed assistance. "Nope, we're fine. Just like to make a grand entrance," I said.

It just so happened that a general had come along on that trip and got the ride of his life. As the airplane came to its all too ungraceful stop, I heard him gulp for air. I glanced back to see that he had turned pale and seemed to have aged a few extra years from the experience.

My spectacular landing skinned a tire, turning it into shredded rubber. I limped the aircraft off the runway to a fixed base operator where they replaced the tire at a cost of $115.

My Singing Debut

Without the ice-melting chemicals we have today, we were at the mercy of Mother Nature. Forced to extend our stay and wait for better runway conditions, we visited a Mexican restaurant that was recommended for an authentic South of the Border meal. The restaurant was interesting in that it didn't have the traditional tables and chairs you'd expect. The setting was more like someone's living room or patio with sofas and a coffee table.

As we worked our way through the appetizers, a woman with a lovely singing voice began serenading us in Spanish. She wore a brightly colored dress with yellow and red flowers, and was accompanied by a man playing a guitar and wearing a sombrero. I recognized many of the songs, one of which was "Bésame Mucho" and began singing along with her in Spanish. When she heard me, she walked over and asked how I knew the words. I didn't feel the need to tell her that Dave Luna had taught me the lyrics to many Latin American songs such as "La Bamba," "Solamente una vez," "La Cucaracha," and "Somos Novios" while we were held in a prison in North Vietnam. I just told her that I learned the song somewhere along the way.

We had a grand time and she invited me to come back anytime. I went back the following day and the day after that. For two weeks, until the ice melted, I roamed the restaurant, singing as if I were Desi Arnaz. What a great gig that was. All I needed were bongos.

Being a test pilot was a hoot and I made it as much fun possible. One assignment took me to Eielson Air Force Base near Fairbanks, Alaska, to test the viscosity of a synthetic hydraulic fluid in extremely cold temperatures. The synthetic hydraulic fluid is less flammable than ordinary petroleum-based fluids and provides a greater survival factor in combat. Oddly enough, it was the hydraulic fluid that caught fire on my aircraft that forced me to radio for help from the KC-135 back in Vietnam.

Eielson Air Force Base was an Alaska Air Command installation under the Strategic Air Command (SAC), and we were Air Force System Command pilots and considered visitors. When we arrived at Eielson, with the maneuver preplanned, we called to the tower that we were on initial approach. Once given the okay to land, the KC-135 pitched out and took spacing ahead. I waited four seconds and then made my turn and together we came in on final. I landed behind him separated by three thousand feet. We must have looked like a sparrow chasing a hawk. We weren't just screwing around; we were test pilots and this was serious stuff.

After securing the airplane, we went to see the airdrome commander to announce that we were there to conduct cold weather tests on our aircraft. He looked up at me and in a most official manner said, "We are Strategic Air Command and we don't fly our aircraft down initial and pitch out and land like that. I don't want to see those two aircraft flying down on initial and landing in formation again," he said. He seemed friendly enough.

"No problem, sir," I said.

Apparently SAC had different rules than the Air Force System Command, and wanting to obey them, the next morning, using the same profile we had when landing the previous day, as the KC-135 rolled down the runway for take-off, I was right on its wing.

Flying the F-4 with a civilian test engineer as my backseater, and Captains Peter "Puck" Larkin as the pilot and Bart Switzer as the co-pilot of the KC-135, we began testing to see how the synthetic hydraulic fluid reacted in the F-4 systems and in the KC-135's larger reservoir.

To test the systems, we cold soaked the airplanes in the freezing -20 degree temperatures for twenty-four hours to allow all the components to freeze up, and then flew two different profiles while keeping the hydraulic fluid coursing through the system. One profile was flying at low altitude with aggressive maneuvers to keep the hydraulic fluid recycling through the aircraft flight controls while the other profile was at high altitude with minimal flight control movement. The more active profile was to see if the fluid would stay viscous enough to support rapid changes in the aircraft's attitude while the second was to determine if the synthetic hydraulic fluid would turn to proverbial molasses and impact the ability to operate the control mechanisms. For the KC-135, that meant flying gentle maneuvers, but for me, piloting the F-4, it meant popping up over trees and chasing moose and sometimes rolling in over Alaskan Pipeline workers and watching them scatter for cover. I could have flown out over the wilderness and done a series of aerobatic maneuvers, but what self-respecting test pilot would want to do that?

After completing our testing and harassing a few moose, we landed and reported to the airdrome commander once again.

"We appreciate that you're here testing your aircraft and know that you're not a part of Strategic Air Command, but the manner in which you made your landings and takeoffs is not the way we operate our aircraft here. I don't want to see anymore formation takeoffs with the KC-135 having an F-4 on its wing."

"All right, sir. No problem," I reassured him with a snappy salute.

Next day, keeping true to our word, we took off with the KC-135 on my wing.

The duty officer seemed a little hot under the collar when we returned. With his fists planted on the counter he said, "Guys, I don't want to see that F-4 on the wing of that KC-135 or that KC-135 of the wing of that F-4 again. In fact, I don't want to see those two airplanes in the same airspace at the same time ever again! Do you understand?" Veins bulged in his forehead as he spoke. I got the impression he was not pleased.

We understood perfectly. We loved a challenge. So, for our final flight, and timing the maneuver, I called for a normal Ground Control Approach at ten miles out, flying under radar control. At twenty-five hundred feet, the KC-135 called for a visual straight in approach under VFR conditions. The tower cleared the KC-135 but didn't see them because they weren't looking for an airplane coming straight in—or nearly straight down, I should say.

The tower saw me and gave me clearance to land and informed me they had my aircraft on a radar approach and I was cleared to land. The KC-135 radioed back that they had a visual with the F-4 on GCA final. The tower told them to continue, although they didn't have them in sight.

The KC-135 called five miles on visual, and I called five miles on GCA.

"AGAR01, cleared visual straight in, maintain visual separation from the F-4 aircraft on GCA final."

The KC-135 radioed back, "I have a visual, maintaining clearance."

"Cleared. You're number two behind an F-4, four miles on GCA final."

"Tally ho," the KC-135 replied.

"The F-4 is about three miles on final," the tower relayed.

They cleared me to land on GCA and cleared the KC-135 to land on visual with instructions to maintain visual separation from the F-4.

I called one mile on final.

"Cleared to land. Maintain visual separation from the fighter," the tower told the KC-135 crew.

The KC-135, with its engine set on idle, full flaps and landing gear down and spoilers up, came in perfectly on my wing. We landed together like a choreographed ballet. It worked out beautifully,

Relief melted across the airdrome commander's face when I announced that we were leaving the next day, but he still got in the last word. "You guys better not take off at the same time."

The next morning, we played by their rules and took off according to regulation. The KC-135 took off and after the appropriate length of time I took off and joined up on the KC-135's wing. We flew out a couple miles, turned and came back around, flew down the runway together, and then dipped our wings in a salute. Like I said, we made flying fun.

The fact that we occasionally broke the rules didn't endear us to the folks at Holloman Air Force Base in New Mexico either. Although Holloman was a test facility and accustomed to various types of testing done in their area, they weren't prepared for Mechenbier and crew.

Our mission at Holloman was to evaluate the effectiveness of a new electronic jamming pod against radars that had been employed by the North Vietnamese to attack American aircraft in Vietnam. The vast open mesa in south central New Mexico provided the perfect landscape for testing where there was no place to hide, giving us the opportunity to test the pod's effectiveness without background terrain to mask the aircraft. At the same time the flat area allowed us to determine if the proximity to the ground enhanced the protection. In short, it legalized buzzing in the name of operational testing.

Using the desert terrain as our test palette, we flew one hundred feet above ground level in the open area, and we also flew over built up or cluttered areas as well, to fully test the pod's capability. That meant flying over the base at five hundred feet at five hundred knots, which grabs the attention of unsuspecting folks on the ground. Among them was Major Bill McGuth, director of operations, who was not dazzled by our brilliant airmanship and didn't buy into the reasoning that we were flying low over his base because it was part of our important test mission. We were called on the carpet and sternly admonished—we were to cease flying over the base in such a manner.

"You do realize that I'm on a test mission," I said.

You could see his wheels turning as he tried to figure out how to handle this test pilot in an orange flight suit standing before him.

"Just what is your mission here?" he asked.

"Sorry, sir. You don't have a need to know."

His jaw tightened and his face turned red, but before he could utter what he was thinking, I said, "You could call Colonel James Abrahamson at Wright-Patterson. He'll tell you we're legitimate."

That pretty much put an end to his fuming.

We left the next day.

Growing a Family

Jerri and I wanted children. For two years we tried to have a child, but nothing happened. It seems that getting clobbered by that Lacrosse ball while at the Academy had given me a traumatic vasectomy and I was told I could never have a kid, so we turned to adopting.

We tried adopting through regular channels within the United States, but the waiting list was long and we were years from the top. Jerri and I had been attending Mary Help of Christians Church where Father Koverman, knowing our desire, told us about an orphanage in South Vietnam and asked if we would be interested in adopting a Vietnamese child. Without hesitation we said yes. Jerri sent in the application to Catholic Charities in Ohio and we were told they had a child for us. That child turned out to be Mahli, who was found in a ditch outside an orphanage in South Vietnam when she was three days old.

In April 1975 while going through the process as part of the American Operation Babylift of Vietnamese orphans, Saigon fell, and the program came under assault when a bunch of "do gooders" made a stink, claiming we were kidnapping children from their parents, and brought a class action law suit against the agency.

We had gotten Mahli in May 1975, but because of the lawsuit we had to endure a two-year waiting period before we could legally adopt her. But once we had her there was no way we were ever going to give her back. Jerri and I had a policy that if anyone knocked on our door, we wouldn't open it unless we knew the person. Jerri kept a suitcase packed for three years should anyone come to take our daughter away. The plan was to take the baby and disappear. Fortunately, that never happened.

Mahli's Vietnamese name was Anh Thi Thu, but we changed it to Mahli Xuan. "Mahli" means flower and "Xuan" means spring, and because she came to us in May and was beautiful, she became our spring flower.

Like potato chips, we couldn't stop with only one child. We wanted a playmate for Mahli, however by that time the war had ended in Vietnam and there was a turnover in their government and adoptions from there weren't allowed.

We contacted the Holt Adoption Agency in Eugene, Oregon, which was working in Thailand and began the process of adoption with them. The agency was slow processing the paperwork and constantly wanted more and more information. They weren't satisfied with just our fingerprints and bank statements, they became invasive wanting personal information: Where were we born? What were the names of our parents? Where were they born? The list of questions seemed endless. Every time there was a change in Thailand's government, which happened often, the new government wanted us to send another set of fingerprints, another bank statement, another letter from our agency, and another recommendation from the state. The progress became so bogged down that I finally got fed up and called the agency and told the representative that either they send us our daughter by the following month or they could make hamburger out of her.

"Oh, Mr. Mechenbier you don't mean that."

"Yes, I do," I said.

The next month we received notification that they had a child for us who had been abandoned and was extremely malnourished. We were told she threw tantrums and bit people. Undaunted, Jerri said, "Just send her to me, I'll take care of her."

Somying Karuchi came to us in September 1978 when she was about twenty-one months old. We took out the "uch," changing her name to Kari. We had gotten her during the summer but there was no word for summer in Thailand, and it was suggested that we use Naron, which translates to, "dry season." We figured since she came potty-trained that was a good choice.

Realizing how much we loved the two girls, and seeing how great a mother Jerri was, we decided to increase the size of our family again. We wanted another Asian child, but because the law wouldn't allow couples to adopt more that two foreign children, we were prevented from pursuing further adoptions. But once the law was changed, thanks to Ohio Congressman Bud Brown, Jerri began the process again. We tried Japan, but that fell through. Then she found a nun who ran an orphanage in South Korea and wrote to her on July 4, which happened to be the same day that Tai was found on a street corner in Seoul. That adoption went more smoothly.

Well almost.

In March 1981 we went to Chicago's O'Hare International Airport, which Jerri began referring to as her delivery room, to pick up our third daughter, Tai Jung-Hee Shin. I had a speaking engagement scheduled that evening in Toledo, Ohio, and knew time would be tight. The plan was to fly to Chicago, get Tai, and then Jerri and the baby would fly back to Dayton while I drove to Toledo for my presentation.

International flights into Chicago were notorious for running late, and that day was no exception. We waited several hours for the arrival and I kept looking at my watch, becoming more and more antsy with each passing minute. Standing at the gate—which you could do in those days—the representative from the agency mentioned that Tai had an aversion to men and would scream and cry whenever a man tried to hold her.

When the plane finally landed, we were instructed to stand back and wait for the baby to be brought to us. It seemed to take forever for the airplane's door to open. Not one to wait patiently, seeing an airport worker standing around looking as though he had little to do, I grabbed his identification badge and pinned it on my jacket. I stomped down the ramp and banged on the door of the airplane. When the door finally opened a female caseworker stood in the doorway holding two children. I told her I was there for Jung-Hee Shin. Against the protest of the adoption agency's representative, I reached out for her. Tai looked at me as if she had been waiting for me and never whimpered. I carried the baby through customs and immigration, got her passport, handed her to Jerri, and took off for Toledo.

We named our third daughter Tai and kept the Jung-Hee for a middle name meaning gentle and happy, which she is. She was malnourished and had a variety of health problems, but with Jerri's nurturing, Tai soon became a healthy, happy kid.

Six years later, after Jerri got rid of all the baby clothes, she got pregnant with our son, Bernhard Charles. After being told I could never produce a child, Bo was a special gift.

We told the girls around Christmas time that we would be having another baby. Mahli, now ten years old, told me that she knew her mom was going to be pregnant because she had been praying for it, and Kari, now eight, asked what country this one was from.

I can't tell you much about the morning Bo was born in July 1985 except that the Cincinnati Reds beat the New York Mets 4-0 the night before with Tom Browning on the mound.

Adios Active Duty

In June 1975 the folks at the Military Personnel Center at Randolph Air Force Base notified me that I was coming up on eleven years of aviation service and could no longer continue flying and needed to find another job. I immediately got on the telephone and called them to explain that during my eleven years of aviation service I had spent six years sitting on my ass in a prison camp in North Vietnam, but that didn't cut ice with them; I had to quit flying.

I was on the 1975 promotion list for major and was offered the opportunity for Professional Military Education at the Armed Forces Staff College, a selection given only to the top 10 percent of those promoted. That would be followed by a tour in the Air Mobility Command at Scott Air Force Base as a command post duty officer. Hardly a plush assignment. I didn't want to sit around sucking my thumb in some office somewhere the rest of my military career. I told them that offer wasn't something I wanted to do. All I wanted was to continue flying, but they were adamant. No more flying. That sucked!

I was currently working eighty hours a month flying an F-4 as a test pilot and they were telling me that I should go to school and not fly as my primary duty. That wasn't going to happen if I could help it. Accepting the promotion meant I would have had a two-year obligation to continue on active duty with no flying, so I opted to resign my regular Air Force commission and said "Adios, it's been nice knowing you."

I was familiar with the Ohio Air National Guard in Springfield and knew James Harrass, Jack Wilson, and Bill Welde, civilian engineers who worked in the Test Wing during the week and wore the rank of major as traditional Air National Guardsmen flying the F-100 on weekends.

While flying a local test mission it was not uncommon on occasion to chase their airplanes in the restricted airspace we shared and have playtime by "bouncing" them, a term used to see if they were willing to make a few maneuvers in mock self-defense.

I called Colonel Joe "Chops" Hunter, who was director of operations for the 178th Tactical Fighter Wing in Springfield and also the hiring authority, and asked if there were any openings for another pilot to fly their F-100s. I arranged an interview, which basically consisted of one question, "What are you doing now?" I told him I was a test pilot at Wright-Patterson flying F-4s. I must have made a good impression because he said, "Welcome aboard." I overlooked the Navy reference.

That afternoon I went back to Wright-Patterson, processed the discharge papers, resigned my regular commission, and said goodbye to eleven years of active duty service. I didn't dislike the Air Force; I wasn't thumbing my nose at it. All I wanted to do was continue flying, and staying on active duty didn't afford me that opportunity.

CHAPTER 19

Air National Guard

DAPPERLY DRESSED IN BLUE SLACKS AND BLUE BLAZER, I walked into base operations that first day and met Master Sergeant Randy Kelly sitting at the operation's desk. He acknowledged me with a "Good morning, sir," and without hesitation assigned me the call sign of "Squire." Before I was shot down, my call sign had been "Chisel." I liked the new call sign, and since being the world's greatest fighter pilot ranked up there with nobility, Squire fit me just fine.

Joining the Ohio Air National Guard where I defended Ohio from Kentucky, Indiana, Lake Erie, Pennsylvania, and West Virginia, with marginal success against Michigan while assigned to 162nd Tactical Fighter Squadron was probably the best move I could have made. As a result, I spent about six hundred hours in the F-100 Super Sabre, affectionately called the "Hun" and eighteen hundred hours in the A-7 Corsair II, referred to as SLUF, Short Little Ugly Fella, a relatively newer aircraft first used in Vietnam.

Being a traditional guardsman, I worked one weekend a month with a two-week drill once a year, plus extra flying days while on deployments.

As a part-time squadron pilot in the Guard, I couldn't fly just one weekend a month and keep my skill level at its peak, so whenever the question came up as to who wanted extra flying time or go TDY and do something exciting, my hand went up. It wasn't unusual to receive a phone call on a Saturday morning saying that someone hadn't shown up, they were ready to brief for a low-level or a gunnery mission, and could I take his place? I always kept a bag packed just in case I got such a call, and living just minutes from the base, I was there almost before hanging up. Being available at a minute's notice allowed me to fly twice to three times a week and often on weekends.

Flying the "Hun"

I checked into the squadron and was introduced to Captain Ed Cole, who was to be my instructor pilot for the F-100. He wasted no time getting me in the air. "Here's your dash-1 manual. Grab your gear, your first flight is at 1300 hours," he said.

I quickly learned that flying the North American built F-100 was quite different than flying the F-4. The F-100 used as a fighter-bomber for missions flown over North Vietnam in the early years of the war also served as MiGCAP escort for the F-105 Thunderchief, as an FAC (Forward Air Controller), and as a Wild Weasel when used for air defense suppression. Flown over South Vietnam, it became the Air Force's primary close air support aircraft.

When taking off in the F-4, with its two-stage turbojet engines, you had to pull the stick back and when you reached flying speed you eased it forward. If you did that in an F-100 you'd be in deep kimchi.

Piloting the F-100, a more traditional aircraft, as you accelerated down the runway and reached flying speed you pulled the stick back and left it there. Being unfamiliar with the aircraft, as I took off, out of habit, before I reached flying speed, I pulled the stick all the way back. Cole screamed and shoved the stick forward. First lesson learned—the Hun was not an F-4 and you couldn't take off that way.

The F-100, built and introduced in 1953, with many miles under her wing, was not the world's most reliable airplane and each had its own unique idiosyncrasies. When cold, the airplane didn't work very well so we always kept at least one spare aircraft operational and ready to fly just in case one malfunctioned.

One F-100 in particular, tail number 971, had a worn out mechanical stop on the throttle and if you weren't careful you could pull it all the way through idle and shut it off. While coming in for a landing following a training mission in Springfield, Captain Ron Brown pulled the throttle back and flamed out on the runway. We had a great time razzing him about his airmanship.

But what goes around comes around, and one day while in Tucson, I drew the short straw and got number 971. Being an instructor pilot, I was tasked to monitor a four-ship exercise. As all four of us pulled away from the chocks, I got only a few feet down the taxiway when I pulled the throttle back beyond the idle detent and shut her down. The crew chief walked up to the airplane and asked what had happened.

"Damn it, I shut it down. Hurry and get some air in the engine," I yelled down to him.

He quickly got the ground air unit, plugged it in, and got the airplane started again. I caught up with the other aircraft before they reached the far end of the runway to do their flight check, hoping they hadn't notice that their instructor had screwed up. Luckily, they didn't.

The J-57 engine on the F-100 was a cast iron rock and whenever there was a compressor stall, caused by the engine not getting enough air, it backfired making a loud noise, and flames shot out the front of the engine. This happened most commonly when going from idle to full afterburner. The policy was that if there was a compressor stall, the pilot was to park the airplane, but this happened so frequently that we had become accustomed to the plane's characteristics and ignored it.

During summer training in Alpena, Michigan, Brigadier General Harry Cochran, the assistant adjutant general, requested an orientation ride. After giving him the fifty-cent tour of what not to touch inside the airplane, I began firing up the engine. I pushed the throttle from idle to full military power until I hit the mechanical stop, pushed the throttle to out board, and then to the afterburner. I was throwing gas in the back end but not matching it up properly from the air-fuel mixture in the front, and because there wasn't enough air going into the engine, I heard, BOOM! The airplane backfired; we had a compressor stall.

I told the general that I had to park the airplane. He asked why.

"Because you're sitting in the back seat and I can't take off now. If I took off and someone saw that I had a compressor stall with a general sitting back there, I'd be in big trouble."

I still kid him about screwing up the flight because he was on board.

During a similar event, while flying a low level exercise over Terre Haute, Indiana, Brigadier General Phil Bouchard, the vice commander of the Aeronautical Systems Center, was in the back seat. Performing such an exercise meant we needed three miles visibility and a fifteen-hundred-foot ceiling in order to fly five hundred feet below the clouds. The weather was marginal, but similar to the weather in Europe that I had become accustomed to while flying deployments there. We had been airborne for twenty-five minutes when the weather fell below minimum.

"Wow, this is neat. I can't see squat," Bouchard said, just before I announced that we had to turn around and go back.

"Why?"

"Because you're a general and I can't fly under these conditions with you sitting back there."

Flying with generals in the back seat was beginning to cramp my style. I'm pleased to say I never presented a similar issue for other pilots when I became a general.

We were often tapped to do flybys for Memorial Day parades and military funerals around the state of Ohio. Flying in a four-plane formation and staying above one thousand feet, we'd fly over the event in the Missing Man formation, regroup, and then fly on to the next venue. To impress the crowds, we'd light the afterburner to make the engine explode with a thunderous roar, but because we were flying an F-100 with its troublesome quirks, sometimes the fire came out the front of the engine. Other times it didn't light at all.

Colonel Robert Preston, commander of the 178th Tactical Fighter Group, happened to be the commodore of the Buckeye Lake Yacht Club, and requested that we fly over the lake and give him and his cronies a show. Our mission was to fly a four-ship down the middle of the lake and make a big impressive boom. As we got over the water, we ignited our afterburners, but only two airplane's afterburners fired. Our spectacular boom sounded more like a balloon popping.

When working a drill weekend we got our regular pay, and if we flew, we also got an additional fifty-dollar flight pay. During one drill weekend in Springfield, I strapped into my F-100 and discovered it wouldn't start properly. At the same time Captain Bill Lynch, who later became a major general and adjutant general for the state of Pennsylvania, climbed into his F-100 and found something was wrong with his aircraft as well. Spying the airplane on ready standby, I leaped out of my airplane and rushed to it. Bill, apparently seeing the same aircraft, started toward it in a full trot, but with my long legs, I beat him by a hair. Bill scowled and gave me the fighter pilot's one finger salute and started to stomp back toward base ops.

Bill was attending law school at Ohio Northern at the time and needed the extra money he made while flying during drill weekends. Upon seeing another F-100, with the front seat already occupied, he trudged back across the ramp like a kid who just had his ice cream cone stolen, knowing he'd have to take the backseat, a position no fighter pilot ever wants, and climbed the ladder into the cockpit. After strapping in, he came on the radio with, "Damn you, Mechenbier, fifty bucks is fifty bucks."

We had an F-100, tail number 730, which was the first supersonic airplane to across the Atlantic Ocean. The pilot, Robbie Risner, a former POW, took the same route to France as Charles Lindbergh. When I found out the airplane's history, even though it was a two-seater, I made that my airplane. It was dubbed the Spirit of St. Louis II and is presently on display in the Pima County Museum in Arizona.

Despite all its faults, flying the F-100 was the best job a fighter pilot could hope for. It's a great looking airplane, and I am truly proud to say I flew it. It didn't have a lot of systems, but it went fast and made a lot of noise, and was darn near impossible to break. As one eighteen-year F-100 veteran pilot, Les Leavoy, once said, "If your AFSC isn't 1115B, flying the F-100, you might as well be shoveling poop in Louisiana."

Flying the A-7

When I first joined the Guard the squadron deployed once a year to a bed-down base at Ramstein Air Base in Germany, where we flew the F-100. Later, in 1982, when we got the long-range A-7s, our bed-down base became RAF Sculthorpe in Norfork, England, where for two weeks we familiarized ourselves with the terrain, flying low-level bombing missions around the Wash, the tidal basin in eastern England, conducting target studies of Eastern bloc nations.

The Guard provided many opportunities for travel; it also allowed more occasions to get into mischief. Flying in England was interesting. Unlike the United States where a pilot was not allowed to come within five hundred feet of a cloud unless flying on radar control, the Brits believe in the "Big Sky Theory" where, if they encountered a cloud, they flew through it. Their thought being that there was little chance of two airplanes being in a cloud at the same time. Today, they have changed their procedures and have the same philosophy as we do—go around if you're not flying on radar.

Another interesting feature of working in England was when flying a low-level, visual course, if you approached a town you were supposed to go around to the right as you would when driving a round-about. Of course checking out all the golf courses like Prestwick, Inverness, and old Saint Andrews in Scotland was interesting, too. Since RAF Leuchars, a British bomber base, was near Saint Andrews, I got a kick out of flying over the golf course to see if I could get someone's attention as they took a swing.

Flying the A-7 while deployed to Loring Air Force Base in Maine just south of the Canadian border, we trained with the Canadian Forward Air Controllers at the air-to-ground school. There, we supported their gunnery range, calling in strikes and dropping ordinance on specified targets for units engaged in combat training. Just as we had done in Vietnam while flying combat missions, we'd fly to an initial point, and the forward air controller would give us a time and heading. Then we'd pop up and the controller would talk us in on the target. They got training and we did, too.

The rules for flying in Canada are quite different than ours. In the United States with a more congested air space, flying is controlled within designated corridors and pilots have smaller areas in which to operate. Canada's gunnery ranges, unlike those in the United States, are in remote areas and have thousands of acres of uncontrolled air space, allowing pilots to fly willy-nilly.

Flying in a less controlled air space with our skill level and propensity for finding ways to amuse ourselves, we made it our personal mission to do what might not otherwise be tolerated under scrutiny of the Federal Aviation Administration. In other words, we let loose and had fun.

The Canadian FAC didn't see many A-7s and asked that when we finished our training would we give them a flyby so they could have a closer look at our airplane. Their training center was nothing more than a trailer with an observation deck perched on top of a parcel of elevated ground. That was an open invitation for naughtiness. With Lieutenant Eric Smith on my wing, we flew down the fire lane and, at the very last second, popped up from between the trees and watched the training center rock and roll. We gave them what they asked for—up close and personal look at our A-7. Smith, who now flies for United Airlines, still talks about our antics as the most fun he ever had.

We didn't have to be flying to cause trouble. Being fighter pilots, we tried to live up to our reputation for irreverence and never missed the opportunity to perpetrate a practical joke while on a TDY assignment.

Eric Smith, Ted Grovatt, Jack Wilson, Ralph Anderson, and myself, with our personalities being the same shade, were able to pull off what would be considered unprofessional, unmilitary, and downright impertinent. Wearing name tags with Dick Gozinya, Jim Nasium, Ben Dover, and Hugh E. Rection printed on them, we reported to the finance office and presented our military ID cards to the clerk for an advance on our pay to help support our drinking habit. At first glance the names appeared legitimate until read out loud. It was hard to keep a straight face as I watched the financial clerk's expression when it dawned on her what the name tags actually said. The nice little civilian lady behind the counter looked at the names on our ID cards, then at the names on our name tags. A quizzical look came across her face. We had her; this was going to be fun. I expected her to become flustered and spit and sputter, but all she said was, "How come the name on your ID card doesn't match your name tag?"

SIUYOA!

Colonel Robert Preston heard about our shenanigans and was not charmed by our capers.

During the briefing before we deployed to Tucson for Snowbird, a training exercise for fair weather flying during winter months, Preston made his position quite clear as he lectured us on how to behave properly. He told us that he wanted us to stop wearing those disrespectful name tags. "I want every one of you to have a name tag with a proper name on it like mine, Robert E. Preston," he said, and pointed to his name tag.

"Yes, sir," we said in unison.

The next day all five of us with varying ranks showed up to collect our advance pay wearing name tags etched with the name, Robert E. Preston. What else would you expect? We believed in following orders.

Although we had a lot of fun, underneath all the craziness we were still professional and proud that we were proficient enough to accomplish our mission even under the most trying conditions.

Flying any type of aircraft has its dangers and while most problems occur in the skies, there are dangers on the ground as well.

In 1987 I was returning from the gunnery range in Indiana when my A-7 began pitching downward. This is only a minor inconvenience in the air, but while landing, can be disastrous. The A-7 has three hydraulic systems: a power control system 1 and 2, and a utility control system with the power control system number 2 as its primary flight control.

Realizing I had a flight control system problem with the elevator, the supervisor of flying, after assessing the circumstances, recommended that I bypass Springfield for the wider and longer runway at Rickenbacker Air Force Base in Columbus. What I didn't know at the time was the left main landing gear brake was locked and as I touched down short of the barrier cable, the wheel skidded across the cable, causing it to bounce against the airplane and miss the tail hook. Instead of being stopped by the cable, I was free rolling down the runway at 130 miles an hour. Because the brake was locked, the airplane left the paved surface and headed for the gully between the two runways, went down one side and up the other, shot airborne, and then bounced to a stop.

With adrenaline pumping, in a nanosecond I hit the latch on the canopy, disconnected the parachute harness Koch fittings, punched the tabs to release my butt from the 30-pound seat survival kit, disconnected the oxygen hose and microphone leads, and vaulted over the side of the airplane, dropping five feet to the ground. When the well being of your fanny is at stake, you can go through the egress routine pretty darn fast.

Standard operating procedures state that if there is ever a fire in the airplane, once on the ground, the pilot is to run off the nose of the airplane because that's where the firefighters are trained to look for you. That makes sense because should there be an explosion the fire will go off the sides or to the rear of the aircraft, leaving the front clear.

Not knowing if the airplane was about to catch on fire or explode, I took off running away from the aircraft, however, a big cloud of dust caused by my bouncing across the ditch raced along with me. As I came out of the dust, I saw the fire trucks barreling toward me. To avoid being flattened, I made a quick right turn, running as fast as I could. I didn't want to be counted among the statistics where pilots survived a ground impact only to be run over by the fire truck.

That was another headline I didn't care to read: "World's Greatest Fighter Pilot Run Over by Fire Truck While Escaping Aircraft."

Jury Duty

I had been home from Vietnam for almost four years when I received a letter from the clerk of courts in Clark County summoning me to jury duty. Included in the letter was a list of questions, one of which was: Have you ever been convicted of a felony? To the North Vietnamese we weren't prisoners of war; we were criminals and had been tried in a civil court in Hanoi and convicted of air piracy. I knew the clerk of courts and decided to have some fun. I answered, "Yes," and under the "Where, When and What" column, I wrote, "Hanoi, North Vietnam, June 1967, convicted of air piracy, Ho, Ho, Ho."

I figured I'd be on jury duty the next week. I lived in Clark County for the next sixteen years and because of my quasi-criminal conviction was never called for jury duty again. By the same token, I can't be elected president of the United States either.

Freedom Flight #159

As I mentioned, I delayed my chance to be a Freedom Flyer because of all the hoops I was expected to hop through. I was already flying with the Test Wing and the Guard and didn't need their official stamp of approval. But unable to resist the lure of the Charging Cheetahs program and the camaraderie with its Dining In, I decided to become part of the club, so in 1979 I flew my F-100 to Randolph. By then 158 men had gone through the program.

Regulation states that returning POW pilots could fly only Visual Flight Rules (VFR), and the day I was to take my flight the weather was below the twenty-five-hundred-foot minimum. The director of operations took his time kicking around whether I should be allowed to go up that day, but Captain

Larry Brown, who was my seeing-eye dog, said, "Let's go." We found some clear air, buzzed some cows and played around, dive-bombing pretend prey before finally landing. I returned to the traditional champagne reception and was drenched with the bubbly, shot the cork at a map of Vietnam on the wall, and hit the bull's-eye. With that, I had officially become a member of the Charging Cheetahs as Freedom Flyer #159.

As my record of attendance at the Charging Cheetahs reunion shows, I am proud to be part of the 560th team. I'll fly and toast with those folks anywhere, any time. Thanks to them for helping share the past and shape the future. Fly Safe!

Grounded

In 1980, following my annual physical, I happened to mention to the flight surgeon in Springfield that I had experienced a few migraine headaches. I said this in passing, thinking I could trust his discretion. I had had migraines since the time I was in prison but they came and went and never bothered me enough to hinder my ability to handle an airplane, but hearing my comment, the flight surgeon looked at me and said, "You're grounded."

I had been flying between 225 and 250 hours a year without any negative effects from the headaches. That many hours equated to twenty hours a month, about the same amount of flying time active duty pilots were getting. Now, I was being told I can't fly. For a fighter pilot, or any type of pilot for that matter, who loved flying as much as I did, that was like telling a kid his trip to Disney World had been cancelled.

When I went to Brooks Air Force Base for my annual physical the next year, there was an annotation in my medical records that I suffered migraines and that I had been grounded. The doctors there concurred and also grounded me. Once the decision was made that I wasn't allowed to fly, it was written in stone.

I should interject that until this time I had been going to Brooks Medical Center for an annual evaluation where the doctors monitored the returning Air Force POW pilots to see when we might become suicidal. It was a real witch-hunt with no desire to help us. This was in stark contrast to the Navy that had developed a program that actually helped their pilots. When we saw how screwed up the Air Force program was and how the Navy aggressively sought congressional funding to include all the POWs, many of us began going to Pensacola to take our annual physicals with the Navy.

Two years later while attending the annual meeting of the Air National Guard Association of the United States in Denver, Colorado, I was relaxing after the meeting at the bar in Trader Vic's and happened to strike up a conversation with Brigadier General Dan Kellogg, the flight surgeon for the Guard. I explained how I had been grounded administratively for something that didn't interfere with my ability to fly and that I had been trying to have my flying status reinstated.

After downing a few scotch and sodas, Dan said the next time I was at the Pentagon I should come see him. Two weeks later, while in Washington, D.C., representing Fairchild Republic (a civilian job I'll talk about later), I paid Dan a visit. He greeted me warmly, and remembering our conversation, turned to his assistant and told him to take care of the paperwork for a waiver. After two years of sitting on my thumbs, Bam! I was back on flying status.

Squadron Commander

In January of 1984 I was promoted to lieutenant colonel, and in July I was selected for the position of squadron commander of the 162nd Tactical Fighter Squadron. This assignment brought with it many new experiences, some expected, others more clearly off the wall and made me realize that being a commander was more than conducting staff meetings, making routine decisions, and pushing papers. I was responsible for the personal and professional conduct of the men and women under my command, but more importantly, I had to set the example and put aside my fighter pilot antics. Well, some of them anyway.

Although I had been present at many change-of-command ceremonies, this was the first time I was the main focus. Three flights stood formation outside the 178th Tactical Fighter Wing hangar as the squadron guidon was passed from Major James Harrass, the out-going commander, who handed it to Colonel Robert Preston, who in turn gave the guidon to me. With the passing of the flag, began another chapter in the Ed Mechenbier story.

As a flying squadron commander, I now wore two hats—one behind a desk and the other in the cockpit. I looked at the job of commander as an additional duty to my flying job, but of course it was the other way around. There were times staff meetings took precedent over flying, but that was all part of the growing up process.

My main duty was to ensure the combat readiness of the squadron by maintaining my pilots' competency to fly missions and drop bombs anytime, anywhere in the world. The job also included making sure the life support section was properly trained, equipped and staffed, and seeing that the intelligence section was ready and able to integrate into larger units and provide mission essentials. My duties didn't stop there. I was responsible for counseling personnel and adjudicating and ruling on issues involving disciplinary problems. Thankfully, I had a great administration staff in Senior Master Sergeant Tony Younce, Master Sergeant Randy Kelly, and Technical Sergeant Joy Fisher, my right hand go-to guys, and gal, who handled the everyday administrative paperwork, including making sure that personnel medical records were up to date and that I didn't have lint on my uniform.

The leadership, the interaction with others, motivational strategies, and the value that I brought to the equation, however large or small, came from my experience at the Air Force Academy. My training there went beyond learning about the Air Force and how to be an officer, it prepared me for the larger leadership role as a squadron commander.

As a young first-year cadet, I was considered pond scum, and my main focus was simply survival. In my second year, I was able to interact with the upperclassmen as I learned how to become a mentor. By my third year, I had the duty of implementing the system on the class two years behind me. In my fourth year, I had the privilege of overseeing the cadets below me to ensure that everyone knew their role and performed as they should. That last year became part of my growth process and framed the basis for my years as a squadron commander.

The Air National Guard has sometimes been referred to as the raggedy ass militia, but I imposed a higher standard by demanding that everybody have a set of blues and by damn they were going to wear them. When a pilot wasn't flying and in a flight suit, he was expected to be in the uniform of the day. Because I expected everyone to adhere to the regulations by wearing the uniform according to regulations, I had to lead by example, and the days of wearing irreverent name tags and a Pancho Villa mustache (which I did grow during my wilder days) were history.

In the seventies and eighties we were being trained and inspected by regular Air Force personnel and were expected to perform to the same standards as the active duty Air Force. Maintaining those standards gave us pride in our unit, respect for ourselves and for each other. We could congratulate ourselves for being the most deployed unit in the Air National Guard. If there was an exercise anywhere in the world, we were confident in our readiness and ability to roll at a moment's notice. It was a pleasure to be part of a unit that had such a forward attitude and proficient skill level.

My leadership style can best be described as an enabler. As the senior officer and commander, the responsibility of a mission fell on my shoulders, but I seldom had time to do all the necessary legwork. That's why it was critical to find the right people for the job.

Whenever there was a task to be done, I scoped out the best people and brought them together, described the goal or end point, assigned each a specific duty, and then asked them what they needed to accomplish the task. After sharing ideas, I took each person's suggestions and formed the plan so they could see the framework in which to operate. I gave them the resources and tools they needed and then stood back and let them go to work.

I never asked anyone to do anything that was stupid or unachievable just because I had some grandiose scheme. Being a common sense thinker, if there was no reason to do something, we didn't do it. When faced with a regulation that set up a roadblock, to say "it can't be done" was not a satisfactory answer. Failure was an acceptable solution only after all avenues had been attempted and it was realized that the project simply could not be achieved. Finally, when the job was completed, I made sure that the people working on the project got the credit due them.

My role as a leader and mentor was, and still is, to motivate others to reach goals they did not realize were achievable and then show them how. When someone says to me, "Oh, I don't think I can do that," I point out that others think they have the potential and ability to do whatever they choose. With high standards, they can achieve most anything. As sports psychologist, Bob Rotella, once said, "People can only be what they know. Show them by example what they can be."

I told my troops that if they didn't believe they were the best at what they did, they never would be. They must have faith in their ability, and a high regard for their own personal capability in order to match wits and skills and talents with other people day in and day out. That usually headed them in the right direction and took them a long way in life.

As a squadron commander, with the counsel and wisdom of others, I also had the opportunity to judge people's qualifications. Because there was very little turnover of personnel in the Guard, I had time to observe the potential in the pilots and select those qualified to become flight commanders and operations officers. I also had the pleasure of picking enlisted personnel to go to pilot training. There were also men I had to take off flying status because they demonstrated a lack of commitment or didn't have the skills to fly without endangering themselves or others. I had a lot of influence in deciding an airman's fate, but I always tried to use that power wisely.

Being a squadron commander had many wonderful moments, but there was one aspect for which I wasn't prepared—working closely with twenty young women, all under the age of twenty-one. Most females in the squadron performed their duties with the utmost professionalism, but there were a few, though proficient in their jobs, who came to me for personal guidance. Their questions often turned my face red. I always had to make sure I didn't cross the line and become too personal, but I couldn't throw them back into the water and tell them to sink or swim. Thankfully, my administration clerk, Joy Fisher, being slightly older and wiser, often came to my rescue. There wasn't a class on resolving female issues at the Academy.

It's been said that many business deals are consummated on the golf course, and for me my love of golf opened many unexpected doors. Even though I was a lieutenant colonel, I was part of the Saturday morning general officer protocol golf. When they didn't have enough generals to fill out a foursome, they included colonels. That's how I happened to be playing one weekend in June 1991 with General Charlie McDonald, commander of the Air Force Logistics Command. We were walking between the fourth and fifth hole when he asked how everything was going.

"Very good, sir. I'm looking forward to retiring," I told him.

"What do you mean retiring?

"I'm coming up on twenty-eight years in the service, which is mandatory retirement for a lieutenant colonel. There are no flying positions for a part-time colonel, and if I can't fly I'm not sure that I want to stay in. I'm going to hang it up and play golf more often."

"Have you ever considered the Air Force Reserve?" he asked.

"No, never gave it any thought."

That was the end of the conversation and we continued on to the next tee.

A couple weeks later I was playing golf with Major General Ed Bracken, who was the Air Force Logistics Command Chief of Staff. We were on the third tee when he said, "Congratulations, Ed."

Thinking he was congratulating me on my retirement, I thanked him and took a swing at the ball.

"It will be a different lifestyle and more time for golf," I said, putting my driver back into my bag.

"What do you mean?"

"Well, what do you mean? Aren't you congratulating me on my pending retirement?"

"Don't you know?"

"Know what?"

"General McDonald transferred you to the Air Force Reserve."

Everyone at Springfield knew, but no one told me. The form 2095, authorizing a transfer, had been filled out and processed. In a period of fifteen days I had gone from a lieutenant colonel ready to retire from the Ohio Air National Guard to a lieutenant colonel ready to meet the 06-promotion board as a Reservist. So, in June of 1991, I said goodbye to the Guard and went to work in the Air Force Reserve.

When I left the 162nd Tactical Fighter Squadron after seven years, I had held that position longer than any commander before me. And what a great ride it was.

CHAPTER 20

Air Force Reserve

BECAUSE GENERAL CHARLES MCDONALD, God love him, made me an offer I couldn't refuse, I was able to stay in the Air Force. Although this new position didn't involve flying, working in the Reserve did open a new chapter in my life and presented new and challenging opportunities.

My first assignment in June 1991 was as a mobilization assistant on the staff of Colonel John Phillips in the Joint Logistics Systems Center where I worked in an industrial manufacturing environment with defense contractors such as Lockheed, Boeing, and McDonnell-Douglas. There is a law that says 50 percent of the overall depot maintenance on an aircraft has to be done at an Air Force facility. Trying to parse out the 50 percent that we would keep and the 50 percent we'd give to the original equipment manufacturer was a precarious balancing act.

Later, as a colonel, I became the mobilization assistant to Major General Rondal Smith, director of Logistics, Air Force Materiel Command. Our mission was to take the best practices of the Air Force and Navy in terms of aircraft depot level maintenance and implement a new electronic-age by taking the pencil out of the process and instituting a computerized system of information technology. In that role I was the eyes and ears of General Smith where I kept track of what each of the depots were doing.

Most of my time was spent traveling to the depots at the various Air Logistics Centers at Hill, Tinker, and Robins Air Force Bases, where they modified, overhauled, and repaired the aircraft. I became knowledgeable in the depot maintenance process, the cost associated with it, the planning and budgeting, and the allocation of parts.

On June 1, 1997, I was at Pensacola being examined during my annual physical when I received a phone call from General Les Lyles telling me that I had been promoted to brigadier general. Although I was authorized to wear the rank, I waited for the formal pinning ceremony before actually wearing the star.

Being promoted to brigadier general was a big deal and the pinning ceremony was pretty neat. General Rondal Smith, along with Jerri, pinned on my first star. General Smith made a speech touting my accomplishments and my ability to assume the higher level of responsibility that accompanied the rank. Had my parents been alive, I'm sure my father would have been sitting there thinking, "That's my boy." I know my mother would have wondered who they were talking about.

The next month I moved to the Aeronautical Systems Center as the mobilization assistant to Lieutenant General Bob Raggio where I became his alter ego, representing him on the Joint Systems Requirement Review Board at promotions, retirements, ribbon cuttings, and civic leader tours. That was the first time I had several hundred reservists whose careers I managed and watched over and groomed for promotion.

Raggio was a pilot of the ugly, twin rotor HH-43 Huskie helicopter, and proudly displayed a large painting of the beast on the wall in his office. As a way to enlighten him on the finer points of good art, one night I snuck into his office and stole, I mean appropriated, the picture and replaced it with a photograph of an A-7. The next morning my name reverberated off the walls of the building.

That man sure had a good set of lungs.

In April 2000 I became the mobilization assistant to the commander at the Air Force Materiel Command working first with General George Babbitt, and then General Les Lyles with much the same duties as I had with General Raggio. All three men were superb officers and I learned a great deal under their leadership. General Lyles really challenged me to grow in the position. He had a vast background in the acquisition process and taught me well. He was a great mentor and friend, professionally and personally.

When General Gregory "Speedy" Martin took command, I had the primary responsibility of overseeing twenty-three hundred Reserve IMAs (Individual Mobilization Augmentees), distributed throughout the thirteen Air Force Materiel Command bases. My main function was to manage the infrastructure that orchestrated, paid, assigned, and justified all those positions. With many Reservists working an average of one hundred days a year, and because there were always more requests for positions than personnel available, I had to ensure that key positions were always filled.

I also became a jack-of-all-trades, doing whatever General Martin needed done. I chaired staff meetings and represented him at off-base functions. I often made official visits to other bases on his behalf, collecting facts and figures at briefings from the various command components, synthesizing that data, and then presented that information to him. I represented him at social and official military functions and traveled with him whenever he needed me. When both the commander and vice commander were gone, I ran staff meetings and monitored routine activities. I never made any earthshaking decisions, but the job was interesting.

Working for General Martin was a joy. He was an operational type guy and brought that mentality to the Air Force Materiel Command where he spent time representing the command to all our users. It was different working with someone who was outwardly focused and who believed his job wasn't so much running the command from day to day and overseeing the responsibilities of his staff, as it was making sure our customers—the war fighters—knew and understood what the Air Force Materiel Command could do, and was doing, for them. He made clear the fact that the Command was very much engaged in the war fight and was an integral part of it.

The Air Force Materiel Command was sometimes thought of as second tier or an additional infrastructure to the whole Air Force process, but General Martin carried the message that we were a part of the larger Air Force. He was recognized as the leader in charge of supporting all weapons systems that were deployed around the world and was responsible for coining the phrase, "Everyday the Air Force goes to war on the back of the Air Force Materiel Command."

Sitting at his elbow gave me the opportunity to witness a fine commander in action and interact with him on a daily basis. On occasion, in his absence, I acted as the chief operating officer. The social and professional contacts I made opened many doors, and the knowledge I gained on how to run a sixty billion dollar enterprise, such as the Air Force Materiel Air Command, was enormous.

There are two theories about how General Martin came to be called "Speedy." One was because he was a former fighter pilot, and the other because he could really talk. Give him a microphone and he was off and running.

The Arleigh House Massacre

On the surface, my job as a mobilization assistant was dull, but there were times that made up for it. The Arleigh House Massacre was one such event. In 2003 the secretary of the Air Force, James Roche, a former Navy captain, displeased with how the acquisition reform was being handled, spit verbal bullets at senior ranking officers during a Joint Air Force Materiel Command/SAF/AQ offsite meeting. Perhaps because he was the new kid on the block, Roche had a refreshing attitude of, "Don't hand me the excuse for something not working or getting accomplished by saying it's always been done that way because I haven't always been here."

During briefings, whenever he asked a question, he expected an answer, and you damn well better have the right one. Around that time we had an issue with depot maintenance on a particular aircraft, and the question arose as to whether we should give the maintenance to a prime contractor or split the work, which in turn brought up the question of acquisition policy.

I was sitting in the third row with the support folks from the Pentagon during the meeting, when Secretary Roche, dissatisfied with the answers he was receiving, began ranting at the senior officers sitting in the first two rows. He ripped them up one side and down the other furious because of their inability to make correct decisions, and told them in no uncertain terms that the changes they made were stupid ones. I sat watching in shock as he tore those men a part.

Later, I was standing in the hallway with Blaze Durante, a powerful civilian involved in the acquisition business at the Pentagon and who controlled the money that goes to the depots. We were chuckling about the slaughter we had just witnessed when Secretary Roche came out of the room, put his arm around Durante's shoulder and asked, "Hey, how did I do in there? Was I too hard on those guys?"

Working with the defense department and trying to adhere to all the Federal Acquisition Regulations (FARs) can be intimidating. I watched a federal employee go head to head with two defense contractors who both wanted to maintain control of the maintenance of a KC-135. They were told rather forcibly that they would cooperate and share the responsibility or they'd see just how ugly the government could be. A threat like that was totally out of line, but sometimes a difficult situation calls for stern measures.

Working as a mobilization assistant, I came to realize that the government is so big and inefficient that it often becomes difficult to keep everything moving, especially if you try to accomplish a job by the book. While working within the confines of regulations, I discovered that it sometimes takes twenty years to make a decision based on a budget that took six years to formulate but could change every two years as money went away. Quite often that decision was made in a small room over a weekend and a pot of coffee by two GS-15s the day before the deadline.

Flying the Big Birds

Because my job didn't involve flying for six years, my feet never touched the rudders of an aircraft. That was a bad time for me. But when the General Officer Flying Indoctrination Program came to my attention, I welcomed the chance to get back into the cockpit.

The program was limited to one general per major air command and being the senior rated officer in the Air Force Materiel Command not on flying status, the opportunity to be part of that program fell to me first.

Since the 445th Airlift Wing was situated across the runway from the AFMC (Air Force Materiel Command) and because I had never flown a big aircraft, I went to see Brigadier General Paul Cooper, the wing commander, and asked if he had any objections if I became one of his pilots. He said, "Sure, go right ahead," and invited me to fly any time I wanted, as did his successors Rusty Moen and Bob Dagnon.

Getting checked out in a cargo aircraft was quite different from flying a jet fighter. The C-141 had many of the same bells and whistles as any other aircraft, but it also had additional buttons, levers, and push-pulls that I was not familiar with. Major Clay Pittman showed me how all the gadgets worked and guided me through the take off and landing procedures until I became proficient at the yoke.

That training period qualified me to fly the C-141. There were a few restrictions; however, I was not allowed to fly operational missions or carry passengers. All I was permitted to do was fly the traffic pattern around the flag-pole twice a month with an instructor pilot in the right seat, so if anything went wrong, I could claim it was his fault.

With my new skills and restrictions in mind, I began hanging around the 356th Airlift Squadron, one of two squadrons within the Wing, every chance I got in hopes of flying. I walked into base ops one afternoon and asked if I could be put on the flight schedule. Major Rick Webster, the scheduling officer, looked up at me and said, "Yes, you and every other jet jockey around there."

He was winding tape around the grip of a golf club and we started talking about the disease we shared in the love of the game. Having supported the National Science Foundation where he flew from Christ Church in New Zealand to Antarctica, he claimed to have hit a golf ball on all seven continents. I really wanted to know this guy.

After several months of flying with adult supervision and playing by the rules, I asked Brigadier General Perry Lamy, director of operations of the AFMC, to whom I was to report my flying time.

"Nobody really," he said.

I ran the same question up the Reserve side to Lieutenant General James Sherrard, commander of the Air Force Reserve Command and asked if anyone kept track of my flying time.

"No, it's just a good deal. It gives you a chance to mingle with one of the units and for them to talk to a general."

With no one giving a rat's patootie what I did or keeping track of my hours, I began flying several times a month, carrying passengers. With many hours recorded in my flight book, I had become as proficient as any other qualified cargo pilot, and I was responsible enough to know that if at anytime I felt I was not confident in my flying skills, I would never put others at risk.

I became such a fixture around the 445th Airlift Wing that even to this day people working there refer to the unit as General Mechenbier's Wing. I just loan it to the guy who happens to be the commander at the time.

Wanting the experience of flying into Baghdad, when the opportunity arose in 2003 to fly operational medevac missions to airlift wounded service members injured in Iraq to Andrews Air Force Base in Maryland, I was first in line. Webster and Colonel Brian Dominguez, 445th Airlift Wing Vice Commander, and Lieutenant Colonel Steve Johnson, squadron commander of the 89th Airlift Squadron, who were always willing to push the envelope, devised a scheme to get me on a flight into Iraq.

The plan was for me to put on a lieutenant colonel's rank and fly down range from Ramstein Air Base in Germany into Iraq, but I got ratted out when someone saw a major general go into base ops and a lieutenant colonel with the same name tag come out. I guess the scheme was slightly flawed. The operations officer was cool about it and didn't make a stink but cautioned that it wasn't prudent for me to attempt a flight into Baghdad. His reasoning was to protect my ass. They didn't want a "high valued target" flying into enemy territory. Most likely the real reason was they didn't want to be buried under the mound of paperwork should anything happen to a general.

When Rick tried to get me on the flight legally by including my name on the manifest as a crew member, that didn't float either. Rules were rules and by damn they weren't doing to allow me on that airplane no matter how crafty Webster was, so I got grounded and spent a few extra days at Ramstein while everyone else made the flight into Baghdad.

With time on my hands, I went to the hospital at Landstuhl and visited the young men and women who were facing surgery or recovering from wounds. I looked in on two Marines who, while deployed to a small African country to support the contingency there, had been bitten by a horrible disease-carrying bug that caused their bodies to swell two times its normal size. They were extremely contagious and put into an isolation unit, and to visit with them, I was required to wear a facemask and gown. Despite their critical condition, all they seemed worried about was getting back to their unit. I found out later they both recovered and did return to duty.

When the crew returned to Ramstein, I once again took the position in the left seat to fly the aircraft back to the United States. When piloting a transport airplane, unlike a single-seat jet fighter, you become keenly aware that you are responsible for not only the aircraft and yourself, but you are also accountable for the crewmembers and passengers, and on this flight, I had fifteen ambulatory and twenty-one litter patients secured in the back.

We were barely in the air when I heard, "Sir, there's a private back here who doesn't believe that a general is flying this aircraft."

I left the controls in the capable hands of Steve Schnell and went to the back of the airplane. Seeing soldiers lying on stretchers in the crude environment of an aircraft was very different from seeing them in a clean hospital room where a trained medical staff was available if anything went wrong. At twenty-five thousand feet in the clouds and an eight-hour flight over the ocean ahead, I suddenly realized the enormous responsibility I had for the care of these patients. I was no longer just a visiting officer in a hospital, I was a pilot entrusted with men's lives. They were dependent on my ability to get them back home safely.

"So where is this troublemaker?" I ask the crew flight nurse.

She pointed to a young man lying on a litter with IVs plugged into his arm. He had a bag of Frito-Lay chips sitting on his chest and was munching on one.

"Who told you that you could bring those aboard my airplane," I said, and grabbed the chips and began eating them.

When he saw the stars on my flight cap his eyes grew large. "Sir. Sir. Sir," he sputtered. I believe the poor guy would have pulled the tubes out of his veins and snapped to attention had he not been strapped down.

The Army and Marines have a different attitude than the Air Force. These men were coming back from battle in Iraq with various injuries but hadn't been given the opportunity to call home. We had a SATCOM (satellite communication system) on board, so I got a 150-foot-long cable, patched it into the front on the airplane, and dragged the phone around to each patient so they could talk to his family.

It was on that return flight back to Andrews Air Force Base in Washington, D.C., that I came close to getting shot down, again. Okay, that may be a slight exaggeration, but it makes for a good tale. As we approached Washington from the northwest, for some reason we were vectored toward the White House. I knew we were nearing the no-fly-zone and began wondering why the control tower wasn't giving new directions. I also began worrying about the F-15s that might be scrambled to shoot us down if we got too closed to the Capitol. Schnell who was in charge of handling the radio called the tower and told them we were closing in on the restricted area. Someone in the tower finally woke up long enough to give us an immediate left turn, out of harms way, before calling out the big guns.

You would think that it's usually a younger, less experienced airman that screws up and gets a unit into trouble, but that's not always the case. One afternoon during a briefing, Colonel Dominguez announced that the Wing had gotten gigged for not having all their reoccurring altitude chamber training accomplished in a timely manner and needed to do a better job.

"No one is above the requirement of getting his physiological training certified, and we have one individual who is a year overdue on his altitude chamber, General," he said, giving me the ole eyeball. Seems that I was the culprit responsible for getting them written up.

Air Shows

Most everything in my life has happened either by accident or the design of others—serendipity if you will. One of my more enjoyable jobs fits that description. I was watching and listening to the male and female anchors at the Dayton Air Fair in 1975 as they attempted to describe the aerobatic maneuvers of the military airplanes.

The dialog went something like this:

Female voice, "Well, Jim, that's an interesting little airplane they're flying, a pretty little blue thing. What do we know about it?"

Male voice, "Well, it's an early model jet."

Female voice, "How far away do you think it is?"

Male voice, "Oh, about four to five miles away would be my guess."

At which point the airplane zoomed past, leaving the crowd astonished. It would have been funny, but as a fighter pilot, the commentary was a disaster with civilian announcers trying to describe the military maneuvers of the Blue Angels.

I had done a couple interviews with Don Wayne, a long-time anchor at the WHIO–TV station, when I first came home, so I knew the people there. I wrote a letter to John Clarke, public relations director, saying that his newscasters' lack of knowledge on military history and aircraft coupled with their inability to explain aerodynamics was hurting the station's credibility as their announcers bungled their way through the air show demonstration. I told him he needed someone standing off to the side feeding them information. I was sure someone at the air base could be made available to give the anchors a basic course in aeronautics and aircraft history so they could do a better job on the air.

About two weeks later I got a contract in the mail with a note, "Sign on the dotted line. You're our guy."

Come time for the next air show in 1976, I showed up expecting to stand on the sidelines and hand notes to the announcers. That's when I was informed that I was to be the on-air military aircraft expert, and found myself sitting alongside Jim Baldridge, the news anchor, and Gil Whitney, the local weatherman, giving the play-by-play action of the Dayton Air Show.

My job was to explain the complicated and technical aerodynamics of flying a jet aircraft. I cobbled together what I called my flight tutorial to teach folks how an airplane worked. I explained that when the pilot pushes the stick forward, the airplane goes lower and the houses grow larger and when he pulls the stick back, the airplane goes higher and the houses get smaller.

Being the voice representing the military perspective was a great experience and I thoroughly enjoyed it. In 1983 I watched members of the Public Affairs Office and the local media scramble for position on the flight line to grab an interview with one of the Thunderbird pilots. The airplane pulled up and the lead pilot, Major Jim Latham, a former POW, climbed out. When he saw me, he came running, and gave me a big hug. We had a mini reunion while the media stood aside cooling their heels.

After announcing the show for several years, Stan Mouse, who had been the general manager of Channel 7 and the founder of some forty-five television stations around the county, suggested that we export the show. We formed a group called the Fly Boys that included Stan and Garrison Mouse, Bob Romond, Larry Kolpek, Larry Bailey, and myself, and began doing telecast beyond Dayton. Our first road show was in Cleveland at Burke Lakefront Airport, followed the next year by Detroit, Boston, Los Angeles, San Francisco, and San Diego. We did six or seven productions a summer working with each city's local talent making them look good. We didn't make much money, but we had a great time creating a four-hour, top-notch show.

Producing such a show takes many talented technicians and plenty of equipment. It was awesome watching Bob Romond handle fourteen television screens at once while directing six or seven videographers to give him the shots he wanted as he made the air show come alive for the television audience.

While in Boston, Bob directed Larry Kolpek, the cameraman, to follow an airplane all the way to the top of a hammerhead stall. At the apex of the climb, just as the pilot ran out of airspeed, Larry zoomed in on the cockpit and caught the pilot's expression as he turned to look in the direction his airplane was headed.

From out of nowhere a woman came running up to me, grabbed me, and gave me a big kiss, telling me how wonderful I was and that she loved me. That was the closest I ever came to having extramarital sex.

"Whoa. Who are and what . . ." I tried to ask.

"You let me see a picture of my husband I've never seen before," she said.

The woman told me that the aerobatic pilot was her husband, Matt Chapman, who Larry had just captured at the top of the stall. In all the years her husband had flown, she had never seen a close-up photograph of him while in the air and she was thrilled to see what he looked like in the cockpit. Bob and Larry did all the work and I netted the delightful benefit of their skill.

We were involved in an air show at Moffett Field near San Francisco where we had to deal with folks who believed they knew how to do our job better than we could. When we showed up at the site, we had everything we needed—trucks, crew, power equipment, cameras, monitors, directors, producers, and caterers. We were a virtual floating TV station, but the fellow in charge kept insisting they had everything under control and would run the production end of the program; they didn't need our help.

"How about the sound equipment?" Bob asked.

"Got it."

"How about the Kyron operator?"

"Yes, he's in place."

"And how about the generators?"

"Got that, too. We even have a backup," he said.

I could see he was determined, and since it appeared they had everything they needed, we backed off and let them handle the production end, while we did the announcing.

About three hours into the telecast everything went black. We had lost power. We switched to the alternate generator. Nothing.

We were off the air for fifteen minutes, and let me tell you, when you are on the air live, fifteen minutes is like a lifetime. It seems *their* generator and *their* backup generator had both run out of gas. And where was the guy responsible for the generators? He was off enjoying the air show.

Apparently, because of some rule about performing on camera in California, I was required to join the Screen Actor's Guild. Someone paid my dues and I became the only general officer in the Air Force to belong to a union.

We were in Cleveland one year doing a show for WKYC Channel 3 with Crystal Bernard as a guest. She was sitting on my right with the news anchor on my left. Crystal was looking at me as she was talking and the anchor started laughing. When we went to commercial Crystal asked, "Why are you laughing? Did I miss something?"

"You were rubbing the general's leg."

"I was?" she said, with a red face. "I knew my hand was doing something, but I didn't know what."

Unlike her television show, *Wings*, where everything was scripted and rehearsed, being on live television made her extremely nervous.

Naturally, over the years I had built a good rapport with the Blue Angels and Thunderbirds pilots and, being in uniform, was frequently able to invite one of the pilots on the set as a guest for an interview. I always had Blue Angel number 8 give his viewpoint of what it was like to fly an A-4 Skyhawk or F/A-18 Hornet. I never blew his cover that number 8 was actually a navigator and I always introduced him as a Blue Angel. They were all superb naval flight officers and I enjoyed their perspective.

One year the Fly Boys became part of the show. To give the audience a close up view of what the Thunderbirds were doing in the air, we formed our own troupe called the WHIO Leader Chickens and performed highly technical and dangerous low-altitude maneuvers on the runway in—what else—golf carts. You laugh, but this was serious stuff.

We mounted our carts and taxied in a tight formation to the departure end of the runway. To open the show we executed a Diamond maneuver, and then from a Wonton Roll went straight into the Arrowhead position, followed by the extremely dangerous Knife Edge Pass with two opposing solos going head to head in a closing speed of five knots. The choreography was simply amazing.

To further entertain the crowd, the Leader Chickens performed the always-popular horizontal bomb burst while number five aircraft, I mean golf cart, skillfully carried out aileron rolls in the center of the runway as the others returned and linked up with the formation from the cardinal points of the compass, a very delicate and intricate maneuver.

Our final demonstration was a pass in review for an awe-struck audience mesmerized by our precision and daring.

As with everything, all good things must come to an end, and in 2002 there was a change in the Dayton Air Show management and the new boss decided, erroneously, that our production was hurting gate sales. They claimed people were staying home and watching the show on TV rather than buying tickets and going to the field to see the performances. What they hadn't taken into consideration was we actually enhanced their revenue by being a four-hour commercial for the Sunday show as we promoted it on Saturday. We had proven throughout the years that folks in San Diego, Los Angeles, Philadelphia, and Boston who wanted to go to air shows went regardless of a preshow or not.

But the new management's narrow-minded view brought an end to the televised air show in Dayton. We did, however, continue the production in other cities until Stan Mouse died. Other factors such as market dynamics and the big networks putting pressure on smaller, local stations to pay network fees for football and baseball games limited the local station's ability to do programming during primetime on Saturday afternoons and the Fly Boys became history.

CHAPTER 21

Working in the Civilian World

LEAVING ACTIVE DUTY IN JULY 1975 and joining the Guard, and later the Reserve, meant I was working only part-time and needed to find a full-time day job. Fortunately, there were people who knew me during my two and a half years at the Test Wing and were willing to recommend me for a civil service job in the Aeronautical Equipment Systems Program Office at Wright-Patterson. There, I became a program manager in the Electronic Warfare SPO (Systems Program Office), affectionately called the Spook World, working for Colonel Buck Dubee, Lieutenant Colonel John Pedjoe, and Mr. Keith Richie. These wonderful gentlemen were willing to give me a job and let me fail with the confidence that I wouldn't.

Colonel Dubee was a good-looking guy despite being a navigator, and carried himself with an earned swagger. While TDY to Eglin Air Force Base in Florida working on the F-4 Tactical Electronic Reconnaissance System, a group of us were sitting in a local haunt enjoying a bit of the spirits, when I saw Dubee in the corner of the bar. I called the waitress over and asked her to take a glass of buttermilk to him with the message that real fighter pilots drink buttermilk. I watched from my perch as she handed him the glass. Dubee looked at the buttermilk, then scanned the room, and with a smirk, stood and ceremoniously chugged the buttermilk, and then yelled, "Mechenbier, where in the . . . are you?"

The Spook World

My first dive into spy technology was with the RC-135 reconnaissance aircraft that flew along the perimeter of the Soviet Union and Warsaw Pact countries looking for and collecting data hidden in the beeps and chirps of radar signals. As the program manager for the project, I worked on a classified signal intelligence collection platform called Compass Quick, a version of the RC-135 that used antenna to detect what the Russians might be up to.

To give you an idea of how covert the program was, the Strategic Air Command owned the aircraft, Air Force Logistics Command equipped the airplanes, and the Tactical Air Command was tasked to operate them. With each agency having a portion of the funding, no one knew where all the money was going, eliminating government scrutiny. A couple hundred thousand dollars here and a couple hundred thousand dollars there was chump change.

I'm still involved in the Blue Devil Program within the Big Safari Program Office at Wright-Patterson doing classified work where we operate several types of aircraft, carrying out unique missions. Some of the aircraft are remote-piloted aircraft; others are propeller-driven corporate aircraft like the MC-12 Huron made by Beechcraft. But the real technological magic is in the collection sensors on board the Beechcraft King Air 90 with its electro optical, infrared, and full motion video cameras. The technology is so advanced that interpreters can analyze a four-by-four-kilometer area of land taken from fifteen thousand feet in the air and detect a pimple on a person's nose. It's amazing stuff.

I could tell you more, but then I'd have to come to your house and kill you.

My next civil service job was working as the multinational program manager of sales of the F-16 Fighting Falcon to Greece and Turkey, in competition against the French designed Mirage F-1.

Knowing I had been a test pilot, James A. Abrahamson, who was now a lieutenant general, asked if I would like to work with him in the F-16 Systems Program Office. I thought about that for all of two seconds before saying, "Well, okay." I requested a transfer from AEW (Airborne Electronic Warfare) and became the government representative to General Dynamics, traveling to air shows around the world while GD attempted to sell the F-16 to countries wanting to upgrade from F-100s and F-4s to a more sophisticated aircraft.

My job was to make sure that General Dynamics presented only authorized information to each individual country in accordance with the International Traffic and Arms Regulations. The best part of the job was that I got to travel to foreign cities I would not have ordinarily visited, such as Farnborough, Paris, and Berlin.

Paris Air Show

One of those trips took me to Le Bourget near Paris, France, where I laid the groundwork for a new air show. During the presentation, my job was to explain the attributes of the F-16 as the airplane strutted its stuff, doing barrel rolls and cloverleaf maneuvers to an audience of military and civilian buyers.

Wanting to provide a comfortable environment for our clients and make a good impression, General Dynamics hosted an extravagant reception at the chalet, a temporary structure built specially for the occasion. With approximately twenty-five hundred square feet of space, the chalet had a huge bank of windows where they could sit and watch the airplanes perform.

After dinner several test pilots; our lovely female hostesses; Zim Zimmerman, director of logistics for General Dynamics at Fort Worth; and I went for a night on the town with a delightful cruise on the Seine and a show at the Moulin Rouge, followed by a grand ole time on the Champs-Élysées. To top off the evening, we stopped at a little café for drinks and listened to jazz. Zimmerman had paid for dinner so I picked up the bar tab, paying with my credit card.

Back at the chalet the next day we were sitting on the veranda when one of our hostesses came up behind me and put her arms around me and said, "Mr. Mechenbier are you rich?"

"No," I said.

"Are you very rich?" she repeated.

"No, I work for the government."

"Well, I think there's a young gentleman here you'd like to meet."

With that, a well-dressed man walked out onto the veranda. He looked vaguely familiar but I couldn't place him.

"Hello, sir. I was your waiter last night at the club. Is this yours?" he asked, and placed my credit card on the table. This nice young man had taken the train all the way from Paris to Le Bourget and bought admission to the show in order to find me and return my credit card. I was astounded. Not many people would take that much care in returning a credit card.

Wanting to repay him for his effort and honesty, I gave him an F-16 cup and my F-16 tie tacks. I offered him tickets to the air show. I even offered money, trying to show my appreciation. All the while the young lady was giggling.

Wondering what was so funny, I asked what was going on.

"Do you know who he is?" she asked.

"Yeah, he was our waiter last night at the club we went to."

"Do you know his name?"

I thought for a moment. "Yes, he told me his name was Rothschild something."

"Yes, Rothschild of *the* Rothschilds. He owns the place."

There I was giving trinkets to a multimillionaire. I'm sure my face turned fire engine red. I can laugh about it today, but at the time I was horribly embarrassed.

Wearing Two Hats

While I was working in the F-16 SPO, Steve Hix, manager of the Dayton office at Watkins-Johnson, tracked me down and offered me a job as a field sales engineer with his company. Since this meant an increase in pay and job title, I accepted the offer and went to work for them in November of 1977.

Watkins-Johnson's main focus was on reconnaissance system development and manufacturing technology programs. As a sales engineer, I had to be technically smart and competent on the aircraft's jamming systems and know what the black box did. With my degree from the Air Force Academy that focused on electrical engineering, I passed myself off as majoring in electrical engineering. Although the Academy didn't have such a degree program, I could speak the language and knew the system's architecture, and managed quite well.

I spent one week a month in Palo Alto, California, where I worked half a day looking at products and reading specs during sales engineer's meetings. The rest of the day I spent on the golf course. It didn't get any better than that. Unless you count meeting Joe DiMaggio at the Half Moon Bay golf club and seeing two young people making whoopee on the beach.

In the middle of June 1978 I got a phone call from Bob Sanatore, the senior executive with Fairchild Republic. He told me they were building the A-10 Thunderbolt aircraft and needed someone who knew the technical components of the airplane and who could also fly it, and asked if I would be interested in going to work for them.

"How much are you offering?" I asked.

"$28K," came his answer. That was a lot on money back then.

I didn't want to leave Watkins-Johnson; I was making twenty-three thousand dollars with many benefits, such as a lot of time on the golf course in Palo Alto. I figured I'd make him a highball offer that he couldn't meet and that would be the end of it.

I asked for $30K.

He didn't blink.

I went to work for Fairchild Republic Company, earning six million frequent flyer miles between Dayton and New York where the main office was located. I worked for some great guys: Bob Dickson, a retired Air Force general and president of the company; Tom Guarino, head of business development; and, of course, Bob Sanatore.

Once the A-10 program was completed, the company's focus turned to the T-46 trainer, intended to replace the T-37 as the primary trainer for the Air Force. With that new program, I was given the title of both marketing director and program manager, responsible for heading the business and international sales program. My main function was to sell the T-46 trainer and the ground attack version, the AT-46, to governments around the world.

I worked with the program from the time the Air Force developed the specifications for the airplane through refinement of the need statement, to the issuance of the Request For Proposal (RFP) to industry, through the Fairchild Republic design process, and ultimately the USAF source selection process. I was in frequent contact with Air Force representatives at the Pentagon, Air Educational and Training Command at Randolph Air Force Base, and the Systems Program Office at WPAFB, which had the task of making the next generation trainer a reality.

I was to have no contact with the Air Force after the proposal was submitted or during the source selection process. However, I played squash every day with Colonel Whitmel B. Swain, the Air Force program manager. We never violated the spirit of the law by talking about the program. We only played squash.

Two days before the award was announced, I received a call from Swain telling me that Fairchild had met or exceeded all agreed upon specifications and had won the contract. He cautioned me that I couldn't tell anyone because once the Air Force selected the contractor, by law the contract had to be approved by the secretary of the Air Force. After that, Congress would be notified, which could take up to seventy-two hours. For the next few days I walked around with a huge smile, knowing we had won the contract.

The day of the great unveiling, prospective buyers stood around waiting in anticipation as "Fanfare for the Common Man" played over the sound system and smoke from dry ice rolled in around the aircraft. The hall echoed with oohs and aahs when the cover was removed revealing the new T-46. There was only one small problem, or should I say fifteen hundred small problems—that first airplane was missing fifteen hundred parts. The missing portions of the aircraft, made of papier-mâché, cardboard, and plywood, filled the gaps and everything unraveled from there.

Even before the great reveal people were getting wind that progress on the aircraft wasn't going as it should. General Larry Skantze, commander of what was then the Air Force Systems Command, was scheduled to tour the plant and review the process. Everyone was prepared for his ten o'clock scheduled visit. Knowing Larry from when he worked at Wright-Patterson, I decided to walk the line early and talk with the men on the floor to make sure everything went well. When I arrived, Larry was already talking to the crew and they were explaining to him what would and wouldn't be on the airplane. Larry caught on quickly that the production was in trouble.

Several months later, during an Air Force Association show in Washington, Fairchild had a booth set up displaying our airplane where we gave a narrative about the contract we had won and how we were going to build the ground attack version of the AT-46. When Larry came through the Republic booth, he smiled and whispered to me, "Are you still with this screwed up company?"

In the end, the company lacked the capacity to build the airplane and it never went into production. We did, however, build and deliver two aircraft that flew and tested well and were capable of flight, but the relationship between the Air Force and Republic was doomed.

I had a wonderful secretary by the name of Ellen Hertlein. Her husband, Bob, and I were sitting in the grillroom at the Walnut Grove Country Club one Saturday morning after a round of golf talking about Ellen. I was telling Bob what Ellen and I had done that week and he was telling me about what he and Jane had done on the weekend. (Her first name is Ellen and her middle name is Jane.) The young man working at the club asked if we realized we were talking about the same woman?

"Oh, we have a great thing going," I said. "I have her during the week and Bob has her on weekends."

"Yeah, we share," Bob said.

That juicy bit of gossip probably ran through the country club for weeks.

Ellen was an easy mark for my brand of humor and I could really make her blush. She had played a critical role in the company, so on a trip to Long Island for a company party to celebrate winning the T-46 contact, I invited her to share in the celebration. As I checked into the hotel, Ellen, standing near me, saw me put my credit card down on the counter and request a single room.

"What about me?" she asked.

"Oh, you want your own room?"

"Okay," I said to the clerk, "give her a separate room, but can you put us at the end of the hall, someplace where we can go between each other's room discreetly?"

Ellen gave me an elbow in the ribs for that remark.

The next morning, as I was checking out Ellen approached the counter. Seeing an opportunity for mischief, I turned to her and said, "Did you remember to go to your room and mess up the bed to make it look as though you slept there?" That comment earned me another elbow jab. Ouch!

Luckily she's a good sport.

During those four years with Fairchild Republic I lived out of a suitcase. I would leave Dayton for New York on Sunday night, drive onto Long Island on Monday morning against the traffic, spend three days at the plant, and then fly down to Washington to tread the halls of the Pentagon. Following meetings there, I would trek up the hill to the Capitol to meet with congressional delegates, and then fly home on Friday night. The weeklong activities were exhausting.

Staying in hotels got old, so after two years I rented a room in Washington from Dick McConn, a fellow Academy graduate, who was also in the international arms trading business. I was amazed to discover that Dick and I had a history going back to our Vietnam days. Dick had been the intelligence officer in Thailand and was working the daily report, compiling the post strike mission summary the day I was shot down. When he saw my name come across his desk, he knew I had been hit. It's a small world after all.

The easiest solution to always being on the road was to move to Long Island. Jerri and I discussed the idea and decided it was too expensive, besides all our family was in the Dayton area and the kids were settled in school.

When Tom Guarino began pressuring me to move to New York, I told him, thinking he'd never meet my demands, that in order for me to move there I wanted three things.

"First, I want a six-digit salary."

He said, "Okay, no problem."

"Second, I want the title of vice president, and third," I said, looking him directly in the eye, "I want the company to provide financing so I can buy a house in New York comparable to my house in Ohio."

He hesitated a moment, pushed his chair back from his desk, folded his arms across his chest, and looked up at the ceiling. I knew I had him; he wasn't going for it.

"Well, okay. We'll do that, too."

Now, I was worried. I raised the stakes even higher.

"And I want a helicopter to pick me up everyday on my front lawn and fly me to the plant because I'm not going to drive the Long Island Expressway everyday."

He looked at me and said, "You really don't want to move do you?"

I don't know what I would have done had he said okay to that final demand, but I got out of moving to New York.

In those days lobbyists weren't registered, and I didn't need an appointment to see a congressman. I spent a lot of time on Capitol Hill looking for funding for our aircraft and enlightening representatives on the advantage of the Fairchild airplane. Walking into the office of Charles Wilson from Texas was like walking into a Playboy mansion with beautiful girls working for him. On Thursday afternoons, I waited along with three or four other contractors as Congressman Dan Sykes from Florida welcomed us into his office with a glass of scotch before asking us what we wanted.

After meeting in the bar of the Watergate Hotel every night at seven o'clock for five years, I could tell you how many light bulbs and tiles were in the ceiling. There, we discussed how we were to going to accomplish the job politically, and then laid out the strategy as to how we would go forward. Discussions often began with the sale of the A-10 aircraft being built by Fairchild Republic, and evolved to the source selection and the T-46.

In April 1983 I moved up in Fairchild Republic to become director of the Manned Weapon Systems and the field offices where I was responsible for developing and executing business activities for the A-10, T/AT-46 and SF-340 aircraft. This new position took me to various parts of the world.

With the weather so damn hot in the Middle East during the summer months everything shut down and everyone with money got away from the stifling heat. Arabs from Egypt, Abu Dhabi, and United Arab Emirates would postpone their meetings until July and we would meet in London. I visited some absolutely magnificent hotel suites where the food and alcohol was out of this world. Arabs don't drink in their own county, but didn't hesitate to down a few while in England and often quoted the expression, "Allah has eyes, but he can't see over the mountain."

Whenever I went into Germany, customs officials would escort me to a small room and ask in a mild form of interrogation why I carried so many different visas to Arab countries. I could have told them I was a spy, but I wanted to be able to return home.

Except for the time when I was young and walked into Roemer's grocery store and asked for a job, I never had to fill out an employment application. Somehow jobs always found me, and going to work for General Electric Aerospace was no exception.

In the fall of 1985 the secretary of the Air Force, Verne Orr, was in town for the Armed Forces Communications Electronic Association dinner at the Convention Center in Dayton and happened to be in the restroom when the manager of the General Electric office came in and asked, "When in the hell are you going to get off GE's back?"

"Who are you?" Orr asked.

The man identified himself and the next day he was out of a job. Two weeks later a headhunter came looking for me and, in January 1986, I became an employee of General Electric, as manager of the eastern region of aerospace business development where I was responsible for the corporate contact with senior USAF officials. My area of responsibility included the eastern half of the continental United States with seventeen office sites with forty-two employees in eleven states. I represented fifteen departments that produced radar gun turrets, tank transmissions, simulators, and space vehicles, plus all other Aerospace Business Group products. This was the first time I was no longer in the business of producing hardware and working as a technical guru. Now, I was a manager of resources and people.

I had people working for me who were twice my age. Most had been with the company many years. By their standard, I'm sure, I was a young whippersnapper in charge of paying their salaries. That was a real learning experience, and I took it seriously. Wanting to hone my management skills, I did a lot of reading and studying on how to better understand the relationship needed to work with these fine gentlemen. Unlike my position as a squadron commander, I wasn't a father figure or a mentor to these men. They knew much more about our customer's needs than I ever would, and I had to gain their respect in order to keep them motivated.

During that time, General Electric Aerospace was being bought by Martin Marietta. Kaye Johnson, who worked for Science Applications International Corporation (SAIC) in Dayton was looking for someone to be the singular face to the customer—a senior person for all SAIC activities. That's when I went to work for them in December 1992 where I held the title of Corporate Vice President for Development, functioning as the senior SAIC executive in the Midwest and representing the company to the USAF and to industry alike.

SAIC had contracts with the U.S. military and Department of Defense as well as private industry, providing information technologies services and solutions. Having run the business development activities for General Electric Aerospace in the eastern half of the country made transitioning to coordinating eight Dayton-based SAIC business units a snap.

Working for civilian companies while still part of the Air Force put me in a unique position. My military duties gave me access to intelligence that Fairchild Republic, General Electric, and SAIC would like to have had. On the other hand, there was information inside my civilian employment that my Air Force family coveted. My employers understood that I led two lives with two full time jobs. Fortunately, I was able to compartmentalize my work and never violated the trust between the two.

I swapped my civilian clothes for a military uniform two and three times a day as I went about fulfilling my obligations as General Mechenbier one minute and Mr. Mechenbier the next. It got to be funny when General Les Lyles would put his hands over my shoulder boards to cover my rank and say that he needed to speak with Mr. Mechenbier for a minute.

I had worked for SAIC twelve years when my boss, Duane Andrews, called me up one day and said, "I don't know who you are. I don't know what you do, but I need for you to know, I don't want you to work for me, but you're going to." I'm speculating that I had been dumped in his lap and he wasn't too pleased at having someone work for him whom he hadn't hand picked. I got my first review in 2009 after working there for eighteen years.

In 2007, wanting more freedom to pursue other activities such as speaking engagements, serving on various boards, and playing golf, I became a part-time consulting employee at SAIC, working twenty-five to thirty hours per week.

Movie Premier of *Return With Honor*

In 2000 a national premier was held at the Smithsonian in Washington, D.C., showing the movie documentary, *Return With Honor*, that told about American fighter pilots held prisoner in North Vietnam. In the film, I, along with twenty-eight other POWs, were interviewed, each giving his perspective on prison life as archival film footage accompanied our dialog.

Although the film had been out for a couple years, Tom Hanks, having been included in the Prologue, wanted to do something to help spread the distribution. At the premier he was introduced and made a few brief remarks. It was a gala affair.

Following the viewing, fifteen movie posters lay on a table in the reception hall for all the attending POWs and Mr. Hanks to sign. I started at one end while Tom started at the other. When we met in the middle, Tom suggested I go back down the line signing his name and he would go back signing mine. After a brief pause, he said, "On second thought, let's just keep going. I can't spell Mechenbier."

Kevin and Mary Jane McManus and five of his beautiful teenage daughters and several of their friends had come along for the party. I stood smiling surrounded by a bevy of ten beautiful young women having a picture taken when Tom walked up and said, "What's going on? I see fighter pilots get all the beautiful women and here I thought they liked actors."

We had a good laugh and he joined us for a group photo. Thus proving a fighter pilot trumps a movie star every time.

Reserve Forces Policy Board

At some point I must have made a good impression on Michael Dominguez, assistant secretary of the Air Force for Manpower in charge of making nominations for the Reserve Forces Policy Board, because he nominated me and I was designated to the board on October 1, 2001. Along with twelve other representatives from all five services, my job was to advise the secretary of defense, Donald Rumsfeld, on matters concerning the utilization of the Reserve components including training, equipment requirements, and the use of troops within the structure. I had the responsibility of seeing that the Reserve component wasn't viewed as a stepchild and given hand-me-down, or broken and obsolete equipment. I was also involved in discussions about the viability of the Reserve component training standards that included the amount of time allotted for troop activation balanced with civilian employment.

When I wasn't attending meetings at the Pentagon, I visited deployed units in Korea and Bosnia to talk with the troops and find out what they thought about their jobs. Did they feel their work was meaningful, or were they sitting around with little to do? Did they feel they were properly trained and equipped to complete the mission? Were their expectations real? We then fed that information back to the secretary of defense.

For the five years that I served on the board we were successful in procuring and integrating adequate, if not state-of-the-art, equipment into the Reserves; that was a huge win.

Forbes Photo Shoot

Many books had been written about the POW experience. One in particular, *Bouncing Back,* written by Geoffrey Norman, contained information about the wine-tasting class I had conducted while in residence at the Hanoi Hilton. Wanting to write an article for *Forbes* magazine about our experience, Norman called and asked if I would care to participate in a photo shoot to accompany the story. He asked me to select a couple guys who had been in the class to take part. I contacted two good friends, Tom Hall from our banana peel dueling days, and Jim Shively, a fellow classmate from the Academy. They eagerly agreed.

We were wined and dined at McCormick & Schmick's, a pretentious establishment in New York that charged sixty dollars just to go through the door. That was the first time I had ever been in such a swanky restaurant. Tuxedo-clad waiters with white napkins draped over their arms made a grand presentation of our food. Our stay at the Waldorf Astoria was equally luxurious. Forbes spared no expense; it was first class all the way.

I had no idea that taking a photograph involved so much preparation. The morning of the shoot, a cab picked us up at our hotel and took us to the studio. Photographers, makeup and wardrobe people, groomers, and set designers scurried around doing their jobs when we arrived. The entire shoot took eight hours. The article turned out quite well with a double-spread photograph of the three of us holding Baccarat crystal wineglasses, all looking like we were born into the upper crust of society.

CHAPTER 22

Fini Fight—Repatriation Flight

THE LOCKHEED C-141 STARLIFTER, tail number 66-0177, dubbed the Hanoi Taxi,[1] *gained notoriety when it made the famous flight to Hanoi, North Vietnam, on February 12, 1973, to return the first of 591*[2] *prisoners of war in what became known as Operation Homecoming.*

During routine maintenance in 2002, the crew chief of the airplane discovered a blue dynamo label with the words "Hanoi Taxi" on the engineer's panel. Recognizing the aircraft's significance and its distinguished service to our country, 66-0177 was refurbished and painted to its original colors, white over grey, and transferred to the 445th Airlift Wing at Wright-Patterson Air Force Base. For several years it became a flying exhibit with plaques, photographs, and documents preserved in the cabin along with names of service members from Ohio taken from rubbings of the Vietnam Veterans Memorial.

With a career lasting almost forty years, and with more than forty thousand flying hours beneath her wing, she was due to be retired, but the Grand Old Lady wasn't to go quietly. She had one more important mission.

Back to Vietnam

In January of 2004 I finally came to the realization that, at the age of sixty-two, it was time to find a permanent coat rack on which to hang my uniform. One Friday night after flying a local mission, I was talking to Rick Webster and told

---

1. The C-141 known as the Hanoi Taxi is now on display at the National Museum of the Air Force in Dayton, Ohio, and stands as a tribute to the men and women who served on her and in the United States Air Force.

2. This number is a calculation provided by Mike McGrath who, after his release from Hanoi, became the collector of all things historic involving the POWs.

him that for my fini-flight I wanted to take three laps around the flagpole, land, and then buy the crew a beer.

I didn't think anymore about it until one afternoon when I went to the squadron and Webster sprung the news that he and Colonel Dominguez, with the approval of the commander, General Rusty Moen, had planned to give me a grand send-off, but no ordinary party with champagne would do. What I expected for my fini-flight and what I got were entirely two different things.

Because I would soon be retiring as the last Vietnam-era POW still in uniform and on flying status, Webster thought it fitting that I should have the opportunity to go back to Vietnam as the aircraft commander, flying the Hanoi Taxi, for a repatriation mission to return the remains believed to be those of men shot down in August 1968. One of the bodies was from an Army UH-1D Huey Helicopter that crashed in Quang Tri Province and the other from an Air Force O-2A Skymaster reconnaissance aircraft downed in Quang Binh Province, South Vietnam.

Whenever remains are found that are believed to be those of a service member from World War II, Korea, or Vietnam they are taken to Hawaii to be identified by the Joint POW/MIA Accounting Command that is comprised of 250 people from the Army, Navy, Air Force, Marines, and, when needed, civilian scientists and doctors.

Knowing that teams in Korea and Vietnam conducted the operation, Webster made several phone calls and asked if they ever requested different units to take part in the process, and would they consider giving one of those flights to the 445th Airlift Wing. He was told, no; all flight assignments were made by the 618th Tanker Airlift Control Center at Scott Air Force Base in Illinois.

He then called the Air Mobility Command Headquarters and the Tanker Airlift Command Center at Scott and told them he could arrange a special flight that would blow the roof and then explained the circumstances, stressing that the flight had to occur before I retired in June. After much arm-twisting they finally agreed and Webster began making arrangements. He phoned the Air Force Reserve Command Headquarters at Robins Air Force Base in Georgia and spoke with the person in charge of handling airlift operations, training, and real world missions, and working closely with the Secretary of State's office and folks at Wright-Patterson, he spearheaded the operation.

Because of the time difference between Ohio and the Far East, Webster spent many late nights making phone calls and sending e-mails to the American Embassy. The fact that I had been considered a criminal and convicted of air piracy in Vietnam caused a roadblock, which created piles of paperwork. But Webster plowed through the muck and got the necessary clearances and visas. It took him four months, including a lot of sleepless nights, to have a flight assigned to the 445th for my benefit.

I had never considered going back to Vietnam. I had no desire to see what the country had become, but knowing the significance of the flight and how excited everyone was and the tremendous effort Webster had made to make it possible, I agreed.

As the day drew near, my excitement grew, not because I was going back to Vietnam, but because I was being given the great honor of flying the Hanoi Taxi and bringing home my fallen comrades in arms. That made the trip special.

Of the twenty-five crewmembers, eight of us were hard-core golfers, and although returning the bodies to American soil was our primary mission, we figured playing a game or two along the way would be a bonus. Not wanting the public to see us carrying golf clubs aboard the aircraft and get the wrong impression, we smuggled our clubs into the back of the airplane before the press arrived. So when everyone showed up for the big send-off, all they saw were smiling airmen in flight suits carrying flight bags aboard.

I was at the controls when we left Dayton on May 22 bound for Hanoi. I had a terrific flight crew: Lieutenant Colonel Steve Johnson; Colonel Mark Caudill; Majors Rick Webster, John Cherry, and Scott Provost; Captain Doug Jankovich; Senior Master Sergeant Frank Laning; Master Sergeants Mike Kalbfleisch, Kevin Steyer, Robert Welshhans Jr., Ken Wright, John Klemack, Jim Grounds, Stan Evans, and Jeff Whitman; Technical Sergeants Tom (Tony) Wilks, Bill Scherer, John Guillaum, Jeff Hurst, and John Cummings; Staff Sergeant Charles Dyer; and Senior Airman Amanda Crooker.

Knowing that when we landed in Hanoi the Vietnamese would confiscate our passports, John "Cash" Cherry became our money man acting as courier in charge of the twenty-five-thousand dollars needed to get us in and out of Vietnam. We also had two large security forces—Ravens, Technical Sergeant Jeremy Shewmake, and Senior Airman Steven Carroll—who carried big guns to protect the cash and the airplane. This was no small operation.

Because our diplomatic clearances were based on our arrival time in Vietnam, the timing of the entire trip was crucial. We had to be in Hanoi on May 28 and didn't want to arrive too early, and certainly not late. Knowing Murphy and his Law that says anything that can go wrong will, and at the most inconvenient time, we built in contingencies for mechanical problems, crew rest, and anything else Murphy might throw at us. We also planned time for golf, of course.

Our first stop was Travis Air Force Base in California. The weather was less than ideal as we came in for landing and the airplane was bouncing around in the air like an old jalopy on a bumpy dirt road. I put cross controls with left wing down into the wind and rolled the airplane onto the runway so smoothly that the guys in the back never knew we had landed. That set the tone for the trip. Smooth sailing all the way.

After checking into our hotel, we headed to Rancho Solano Golf Course in Fairfield where we found rolling hills flanking the fairways and gorgeous greens and enough water hazards to make the game a challenge.

All the golfers on the flight had a nickname. Steve Johnson's label was "Ish," a handle he acquired for obvious reasons. "We'll meet around twelvish," he'd say, or "We'll leave about four-thirtyish." Johnson was our official scorekeeper and made it his mission to track everyone's score on a high-tech NASA computer gizmo. His system was complicated, and just when we began to understand how he calculated the scores he'd revise his method. We played the course with our scores being mathematically manipulated as we got points for par, bogies, birdies and eagles, who we played with, or whether we teed off from the forward or back tees. The system was designed to keep everyone competitive and encouraged trash talk, of which Doug Jankovich was the master, earning him the pet name, "Blowhole."

Scott Provost got the nickname "Sensei" (the teacher) because he was a good golfer and had the ability to help others with their swing. We made him our resident Pro for the trip. Mark Caudill was often in need of help with the basics, and Sensei was always there with advice. The two had a competition going between them. Whenever Caudill hit a better shot, Provost took it as a challenge to his manhood. They made golfing fun.

Webster, who got the nickname "Bull" from his wife who says he does things in a bulldozer fashion, was always on the lookout for the latest gadget for improving his golf score. If there was a product or thingamabob on the market that guaranteed to improve his swing and lower his score, he bought it. Even with all his fancy toys, his first shot off the tee hooked a low screamer into a ditch.

And so began the Pacific Return Golf Tour.

We left the following morning for Hickam Air Force Base in Hawaii. Because there was no lodging available, we had to endure the inconvenience of being assigned off-base quarters at the Waikiki Outrigger on the Beach where we spent two days, playing golf, and enjoying the local libations served in a coconut. This was becoming a real hardship mission.

We said goodbye to Hawaii the next day and headed to Andersen Air Force Base in Guam. As we approached we ran into a thunderstorm with horrendous hail and turbulence. The crew found no obvious damage to the airplane until we landed and discovered the airplane had blown a hydraulic line, which meant down time, which translated into more time on the links.

We spent the two days in Guam playing golf at Mangilao Golf Course, often called the "Jewel of the Pacific" while the airplane was being repaired. The par 72 course could best be described as a garden surrounded by a golf course. The clubhouse sits on a cliff overlooking a spectacular view of the Pacific Ocean. Imagine playing a course and having a warm tropical Pacific breeze follow you on each hole. Although a difficult course, it was a thrill to play.

With the C-141 air worthy again, our next stop was U-Tapao Air Base in Thailand. Considering the critical timing needed for going through each air defense identification zone, we figured we could best control our entry time into Hanoi by flying directly from Thailand, eliminating the possibility of arriving too soon or too late. Another reason we used Thailand was the need for fuel. Not knowing its integrity or what the Vietnamese would charge for gasoline, we weren't about to take on fuel in Vietnam.

Webster made arrangements for us to stay at the Hard Rock Hotel in downtown Pattaya City. As we checked in, a young lady appeared with a tray of beer. Not wanting to be seen drinking in uniform in public, we politely refused. But when the manager of the hotel appeared and recognized Webster from a previous visit, and after much hand shaking and back slapping, to avoid causing an international incident and ruin Rick's reputation, we accepted the beer.

With the added downtime, we hit the links once again. If Mangilao Golf Course was beautiful, Laem Chabang International Country Club was out of this world. Laem Chabang is actually three courses in one called the Lakes, Mountain, and Valley Courses. The Lakes and Valley links wind along rippling rivers while Mountain holds a fantastic view of the surrounding area. There is no way to fully describe the beauty of the course that would do it justice; it was just that fantastic.

Each of us had a female caddy. Cash had the unfair advantage of having a young lady who spoke understandable English and knew the course. I, on the other hand, had a caddy that didn't know the difference between left and right. In describing the lay of the green, she'd point left and say, "Break right." She was interesting and fun, but not very helpful.

After finishing nine holes, we stopped at the snack bar for refreshments. I was standing at the counter getting a bottle of water when someone yelled, "Look out for the snake!" Looking down I spotted a four-foot cobra slithering at my feet. I jumped, and probably yelled, although I would never admit it. The eight female caddies scurried out of sight. The two servers ducked behind the counter and six American golfers hopped around screaming like little girls.

Tony Wilks, who we called "TW" or "T-Dub," was the only one of us who didn't carry on as though he were about to be devoured by the thing. With his Odyssey putter, he calmly picked up the cobra, carried it across the fairway and threw it into the water. Seems that Tony's father owned a rattlesnake ranch in Texas and to Tony snakes were as common as prairie dogs.

After the cobra incident, we still had snakes on our mind when Mark Caudill approached the next tee. I was in a golf cart slightly ahead and off to the left as he addressed his ball. We were far enough away to not be in any danger of getting hit by an errant ball, but as a precaution and knowing that he wasn't the greatest golfer, I got out and stood behind the cart using it as a shield, just in case. Caudill's shot hooked left, ran under the cart, barely missing my feet. Because the hook, and how the ball rolled, reminded us of a snake, we dubbed Caudill, "Cobra."

The next morning, in order to reach Hanoi at the designated hour, we had a hard takeoff time out of U-Tapao. We flew across Bangkok to Vientiane, Laos, and then crossed the northwestern border into Vietnam, reaching Hanoi two and a half hours later. Along the route, we encountered heavy weather, but as soon as we reached Vietnam and punched through the soup, the sky opened up with sunshine and cumulus clouds.

Although Johnson had been sitting in the right seat most of the journey, Webster requested that position as we flew into Hanoi. While Webster handled the communications, I acted as tour guide, pointing out landmarks, giving a running commentary on the topography. Having flown over the country, albeit many years ago, I still remembered the terrain, and knew every bend in the Red River.

Thirty-seven years later I was once again flying Route Package VI, but this time there were no missiles coming up to greet me. I pointed out the mountain chain called Thud Ridge, and Phantom Ridge where an old SAM site called Lead 110 had given us a hard time. Nearing Hanoi, I noticed bomb craters still dotted the landscape and the Paul Doumer Bridge still spanned the Red River. I was running on adrenaline, seeing and remembering it all.

We had left Thailand with a full load of fuel, but because landing on a relatively short runway with an airplane loaded with so much gas made landing dicey, and because Rick was more familiar with the C-141, I told him if he wanted to make the landing I'd understand.

He said, "Press on, Boss. You're in command."

I had no flood of emotions when we landed in Hanoi. Maybe others might have felt something, but being the stoic that I am, I stayed focused, concentrating on the mission at hand. That's when I fully realized that I had moved on.

As we taxied in, I was amazed to see MiG fighter aircraft in bunkers alongside the runway. In stark contrast to the countryside that seemed to have remained the same for the last two hundred years, billboards now surrounded the airport touting the commercialism of Vietnam. Ads for cigarettes, automobiles, and beer were just a few of the products that you wouldn't expect to see on sale in a Communist country.

I was directed to park the airplane in the boondocks far away from the terminal. We sat sweating in the airplane while John Cherry, protected by the Ravens, went into the terminal office to attend to the protocol of presenting our passports and paying the fee associated with our landing. Negotiations to secure our safe passage in and out of the country and to get our passports back cost us only six thousand dollars, which was low compared to what we had expected.

While Cherry bargained with the Vietnamese, we used the SATCOM radio to call home and talked with our spouses. Several of us got out of the airplane and walked around, stretching our legs for a few minutes while two Vietnamese soldiers followed close behind to make sure we didn't stray. Just like old times.

My status as a former POW returning to Vietnam presented a unique opportunity for the news media. Scores of reporters from the international press, including those from Hanoi newspapers, stood ready with cameras, microphones, and tape recorders to record the event.

Just like in the Philippines, when I first came home I was asked to give an interview to the media. A public affairs officer was there to clarify everything I said to make sure my wording fit with the diplomatic lingo. "What the general meant to say was . . ." and then he would reinvent my words. I found that amusing.

For the most part the reporters did their job in a professional manner cognizant of the solemn occasion. I can't say that about one National Public Radio (NPR) reporter, however. Up until that time I had never had anyone, with the exception of Jane Fonda, who called us warmongers, try to humiliate me or call into question my integrity for having served in that war, and I have talked with a lot of people who had differing opinions about our involvement. That NPR reporter cornered me twice and asked if I would rather be in Iraq bombing children than in Vietnam. I walked away without answering.

As an aside, and to make my position clear, I feel Jane Fonda is a traitor by aiding the enemy. She put men's lives in danger, yet she still maintains she was right and the rest of the world was wrong. I think she is an absolute embarrassment to our country.

When the time came for the repatriation ceremony, some flunky escorted us to the ceremonial area. I saw no ranking Vietnamese officials present as we stood for an hour in the sweltering heat, along with the American Ambassador to Vietnam, the military attaché, and the head of the Joint POW/MIA Recovery Team from Hanoi.

A specially trained U.S. military honor guard conducted the formal repatriation ceremony in a manner befitting the occasion. Meanwhile sweat trickled down my back and soaked my flight suit. During the elaborate service, two small boxes were placed on a table as the first step in their long journey home. The remains were then transferred to two aluminum caskets and reverently draped with the American flag, and then slowly carried onto our aircraft. Aside from the blistering heat, the ceremony went well considering the political environment.

Once again I was at the controls as we lifted off from Hanoi and headed east. I called the crew's attention to the railroad yard that I had bombed the day I was shot down, and when we reached Hai Phong and turned right toward Cat Hai Island, I pointed out where I was almost shot down while flying in the back of the Navy COD.

We hugged the coastline south to Da Nang before turning toward the South China Sea near Cam Ranh Bay. As we left Vietnamese airspace heading to Guam, in a sobering tradition Webster called, "Feet wet. Welcome home."

I flashed back to the day of my repatriation in 1973 when those same words were spoken as our C-141 left Vietnam soil. I still become teary-eyed and choke up thinking about it.

Once over the Pacific, we headed back to Andersen Air Force Base. Even after the four-hour flight, my clothes were still damp and sticking to me when we arrived in Guam.

As we taxied to base ops, I was awestruck to see a large group of base personnel, both military and civilian, of every rank and every branch of service, their spouses and children standing on the tarmac. There must have been three hundred folks waiting there. It struck me that all those people, who could have been enjoying a Saturday afternoon of shopping or time at the beach, cared enough to pay tribute to servicemen who died years before many of them had been born. Seeing that crowd put everything in perspective, that while we were making the most of the trip, having fun and playing golf, others stopped what they were doing to honor these men.

While the crew was busy buttoning up the aircraft for the night, I wandered to the back of the aircraft. From where I stood, I saw dim light shining on the caskets. The only sound was that of the hot metal of the aircraft popping as it began to settle and cool. No one was around and the coffins looked forsaken.

I walked over to them and tapped GBU (God bless you).

The next morning, military personnel lined the insides of the airplane awaiting the ceremony. The back of the aircraft was opened making the two flag-draped caskets visible to the public. In keeping with the solemn occasion, a chaplain said a prayer and the base commander and I placed a wreath on the caskets as a bugler played "Taps."

In an interview later, I remember saying, "I am very honored to be part of this great ceremony today where we recognize that Americans missing in action during the Vietnam War have finally returned to American soil. This tells today's soldier, sailor, airman, and marine that no matter what your fate, you are not written off; you are precious; we will come for you."

We left Guam and headed for Hawaii for the ceremony that was scheduled for June 2 at Hickam Air Force Base.

After making the takeoff, and with the long eight-hour flight ahead, I relinquished control to Webster and Johnson, and went upstairs to the crew area, climbed into a bunk, and went to sleep.

Sometime later, I awoke to the smell of smoke and the sound of the fire alarm. An oxygen mask dangling in front of me. I strapped on my mask and went back to sleep. I figured there was nothing I could do, and Webster and Johnson were capable of handling whatever was happening.

Please note, with the possible exception of running out of fuel or getting shot at, a fire is about the worst emergency you can have aboard an aircraft. And being over the middle of the ocean at night with the only land more than an hour away made the situation even more perilous.

Webster told me later that smoke had suddenly begun pluming from the center console, raising the possibility of a complete electrical failure. Regulations states you must run the checklist if you can't identify the origin of an emergency, but this situation called for a quick decision. Johnson and Webster, who always had a little good-natured friction between them, had a difference of opinion as to what to do. Johnson wanted to turn off all the generators and put the airplane on battery until the source of the problem could be found. Webster being a take action type of guy, (the bull, remember) didn't want to take time fooling with an emergency procedure list.

Although Webster couldn't identify the cause, he knew where the smoke was coming from and chose to bypass the checklist. Tony Wilks, the flight engineer, crawled down into the hole and took a crash axe to the bottom of the panel and ripped it open to discover the soldering on one of the lights had broken loose and was arcing off the metal. He taped the wire and averted what could have become a major disaster. Ish, Bull, and the rest of flight crew deserve an at-a-boy for their fast action.

After all the excitement, I woke up refreshed and ready to make the landing.

Exactly as our careful mission planning had predicted, we arrived two days early at Hickam. I can't say for certain, but I would bet that Webster and Cherry had something to do with our early arrival and had a tee time already scheduled.

Not wanting the press to know we were there, we "hid" the airplane and tried to disappear for a couple days. Unfortunately, the media sniffed us out almost immediately. The interviews I had given in Hanoi and Guam had appeared in *Stars and Stripes* and by the time we arrived in Hawaii everyone knew about the mission and the press was all over us. Because I was a former Vietnam POW and the oldest Air Force general still in uniform on flying status, who happened to be flying the Hanoi Taxi, the event sparked quite a flurry of excitement. All this attention was disconcerting. I wanted the focus to be on the men we were bringing home, not me.

Being the pilot and a general officer, I was asked, once again, to speak to the media. Reporters and photographers crowed close as I repeated the message I gave in Guam, "I am very honored to be part of this great ceremony today where we recognize that American servicemen missing in action during the Korean and Vietnam Wars have finally arrived back on American soil."

I then continued, "The time, talent, and resources devoted to repatriating our MIAs tells the world that we care about our people. That we can bring closure to families who have waited, wondering all these years about their loved ones' fate is an honor, a privilege, and is personally very gratifying because we are backing up what it says on the POW/MIA flag—You Are Not Forgotten."

The group commander greeted us and, after a lot of saluting and handshaking, handed me the keys to a staff car and a room in the Distinguished Visiting Officer's Quarters. I looked at him and then at my guys and asked what he was going to do with my crew. "Oh, General, we have so many airplanes and crews going through here that we can't house everyone on base. Your crew has to stay downtown."

"Are you going to give my crew vehicles?"

"No, sir, we can't give them vehicles either. We just don't have them."

"So, just where are you going to put my crew?"

"We can put them up in the same place you were before at the Waikiki Outrigger."

Let's see. DVOQ (Distinguished Visiting Officers Quarters)? Waikiki Outrigger? It was a no brainer. I handed back the keys and told him thanks, but no thanks, I was staying with my crew.

Cherry explained it was necessary that we stay together as a crew, and needed vehicles large enough to carry everyone, but the colonel wouldn't budge, claiming there simply was no other transportation available.

Not accepting that answer, Cherry announced that he was going to the transportation office to see what could be worked out. I followed with the colonel on my heels. After a brief discussion on the ramp with the Wing commander and protocol officer, someone pulled vehicles out of the air and we ended up with two staff cars and two mini vans. I suspect that someone was left to walk to his office that day.

As we left the tarmac headed for the hotel with Provost at the wheel, a security police car with flashing red lights pulled us over. A young flight line security policeman came up to the car and said, "Sir, did you see the red markings on the tarmac restricting driving in front of the fire house?"

"I did not drive across the red line. I did not go within 150 feet of the fire house," Provost said.

Perhaps Provost's adamant denial must have satisfied the young man, because he nodded and said, "Well, sir, be more careful," and let us go.

We had driven about two hundred feet when the same security car with the same red flashing lights, the same airman stopped us again.

"Sir, you drove across the red line again."

Provost told him that he had not driven across the red line, which started a verbal battle of "You did." "No, I didn't."

The discussion was becoming heated and personal when I leaned forward and asked the airman to call his supervisor. "Obviously we have a difference of opinion here that needs to be resolved," I said.

We sat there in the middle of the tarmac until the NCOIC (Noncommissioned Officer in Charge) of the duty section arrived. After we explained the situation, he said he would talk with the airman and told us not to sweat it. "I'll take care of it," he said.

With everything settled and tempers calmed, Provost looked at the sergeant and said, "Is there any . . . way I can have him stop me before I get out that gate, 150 feet away?"

I wanted to burst out laughing, but all I could do was sit there with a straight face.

Our early arrival gave us time for several rounds of golf at the Waiheke Golf Club and at the Schofield Barracks military course. We also played the Kaneohe Klipper Marine Golf Course where the thirteenth hole is arguably the prettiest in all the military systems. The course goes along the coast at a rather high elevation dropping down to the beach.

On June 1, on the tenth hole at Kaneohe, having hidden a set of master sergeant's stripes in my golf bag, I had the honor of promoting Tony Wilks. Tradition calls for tape to be placed on the back of the stripes and then stuck in place on the sleeve of the uniform with a punch. I positioned the stripes on Tony's sleeve and then gave them a hard whop.

My intention was to impair his golf game.

The day of the ceremony, as before in Guam, I was astounded to see hundreds of people gathered on the tarmac to pay their respects.

Before the ceremony began, I saw a group of Korean War veterans standing along the sidewalk with an honor guard carrying POW/MIA flags and the Vietnam and Korean Veterans of America flag. I walked over and struck up a conversation and discovered that they never miss a repatriation ceremony. I was struck by their dedication, respect, and patriotism.

Webster told me later what he had witnessed while I was talking to the media that morning. In the back of the airplane, he noticed one of the Korean veterans walk by each casket, tap his ring on the lid, and then bend close to the casket and whisper, "Welcome home, Brother." Hearing about that single act made me realize just how much we are a band of brothers.

Standing outside the 15th Airlift Wing Base Operations building with our part of the mission complete, I found myself a thousand miles away, lost in thought. There we were returning men who didn't have the privilege of a Homecoming as I had had thirty years earlier. Now, we were honoring them as if they were actually there. Knowing how close I came to being one of those bodies in a box, I felt a deep sense of gratitude that by the grace of God and two seconds I had been spared such a fate.

Even after three and a half decades, authorities still receive reports every year of Americans missing or unaccounted for in Southeast Asia. The staggering figure reaches into the thousands. Regrettably, their remains may never be found, but until every last man and woman is returned to American soil, we have the duty to keep searching and bring them home so they may have the privilege of a Homecoming.

After the two caskets were carried off the Starlifter, we stood at attention while the bodies of nineteen men from the Korean War, missing more than fifty years, were off-loaded, two by two, in flag-draped caskets from a C-17 Globemaster and carried past a joint honor guard to be placed aboard our C-141 for a brief ceremony before being taken for further identification.

On our last night in Hawaii we played one final game of golf on a lighted, nine-hole par three course. We made arrangements to be the last group out with the intent of having fun while playing by the Mechenbier Combat Rules of Golf where we could wave our arms or a golf club in a player's face, yell, whistle, and scream. The only rule was we couldn't make physical contact with the man on the tee as he took a swing, and belly poking was verboten.

Despite "Ish's" scoring method, I won.

As I climbed into the cockpit, the next morning for the nine-hour, nonstop flight back to Dayton, I realized that having the honor of flying the mission into Hanoi to bring home those bodies was the single most emotional experience of my forty-four years of military service. I am grateful for those crazy guys who made the trip possible.

Jankovich, who was the base ops point man and who coordinated the timing of the flights to guarantee our on-time arrival and departures, left us at Hickam and Lieutenant Colonel Bill Gorzinski took over and stayed in touch by satellite phone to ensure our arrival time for Dayton was on schedule. Adhering to tight arrival and takeoff times is not unusual but I sensed an undercurrent of excitement in the crew's behavior as we neared the continental United States. I discounted it as normal anxiousness of getting home and thought no more about it.

The entire trip took thirteen days and fourteen thousand miles. I was looking forward to returning home to some peace and quiet. Throughout the flight our call sign had been "Reach," the generic call sign for airlifts, but with our mission completed, as we flew over Minneapolis Center, our call sign was changed to "Freedom 159" indicating the end of the repatriation flight and the beginning of my fini-flight.

I was feeling pretty damn spunky as we approached Wright-Patterson. Instead of a straight in approach, I flew an overhead pattern, made a tactical forty-degree bank onto final, landed on runway 05L, and then began taxiing toward the ramp of the 445th Airlift Wing Headquarters.

"Stay to the right," Webster told me.

"No, left. I know where I'm supposed to go," I told him, but he insisted that I turn right. When the "Follow Me" truck began guiding me away from the 445th ramp toward base ops, I began wondering if anyone knew what they were doing.

As I taxied the airplane in front of base operations, I saw the reason for the redirection. Fire trucks lined both sides of the taxiway and had formed a water arch. Beyond the symbolic fini-flight tradition stood a crowd of more than five hundred wonderful people, many waving banners, while the U.S. Air Force Band of Flight played Sousa marches. I knew instantly why the timing was so critical coming across the county; Webster, Dominguez, and Johnson had been up to their shenanigans again. They had to get me back to Dayton on time to celebrate my fini-flight.

Webster kept me busy in the front of the airplane while the crew scrambled around trying to get a case of champagne out the back door. As soon as I stepped off the airplane, corks began popping and the crew, taking careful aim, soaked me down thoroughly with champagne. Then Jerri got into the act with the fire truck's hose and watered me down. I ended up with an inch of booze and water in my boots as I sloshed around thanking everyone for this awesome tribute.

General Martin, Lieutenant General Richard Reynolds, General Bill Looney, and retired General George Rhodes made light-hearted speeches, touting my service and took jabs at my golf score. Keeping with my philosophy that a short speech is the best speech, I thanked everyone, especially the crew, and talked about the purpose of the trip and its true meaning.

With my fini-flight complete, I was one tired and wet puppy, but happy.

CHAPTER 23

Saying Goodbye to the Air Force

THE FLIGHT TO AND FROM VIETNAM was supposed to be my fini-flight, and for all practical purposes it was. But a few days before I retired, I made a second final flight in the left seat. I had mentioned to Webster that I still hadn't taken my three laps around the flagpole and that I would like to have a civilian Type Rating certification in the L-400 aircraft. Webster told me to show up at the squadron and he'd take care of everything.

The testing included a written examination, so I went to the FAA office in Columbus, took the test, and then came back to Wright-Patterson for the flight evaluation check ride. My son, Bo, wanted to fly with me on my last flight, but because regulations state that family members cannot fly together, he was not permitted to go along. That afternoon I showed up at the squadron, wearing my flight suit and name tag; and a guy looking very much like my son showed up with a name tag bearing a bogus name.

Wonder where he got that idea?

Along with Bo, Steve Schnell, the certified FAA examiner, and Steve Johnson, I flew for about an hour around the local area between Cincinnati to Columbus and back to Dayton using the call sign "Freedom 159." Much of the testing involved handling emergencies, such as discovering a fire aboard the aircraft. I had slept through that class on the way back from Vietnam, but still did okay on the test. As I flew through each control tower, the controllers radioed back with, "Good luck, sir" and "Congratulations, sir." Apparently someone had told the tower operators this was my last flight.

I finished the three laps around the flagpole, made a perfect landing, and taxied the aircraft back to base ops. When I landed, knowing that it was my final flight—really my last flight—I literally cried. Yeah, I'm an old softy in my old age. Then as promised I bought every one a beer . . . or three.

Retirement

Even after forty-four years of military service, I wasn't ready for a golden parachute. The only reason I retired was because regulations said that at the mandatory age of sixty-two, I had to.

Having known General Martin, my commanding officer, for only six months, I didn't know if it was appropriate to ask him to conduct my retirement ceremony or if I should ask someone who knew me longer.

The situation resolved itself in January as General Martin and I were leaving Edwards Air Force Base after making a tour of the flight test center there. Before we took off, General Martin said he needed to make a pit stop and then headed to base operations. Protocol states that the highest-ranking officer always enters a vehicle last. I climbed aboard to wait for General Martin, and had just settled into my seat when I saw the steward close the doors and heard the pilot started the engine.

"Hey, wait a minute," I yelled. "You're about to leave without the main man here. We need to wait for the general."

She looked at me, and then at the two stars on my shoulder, and said, "Sir, when we see a general come on board, we move out."

"Well, you missed the four star," I said.

Once General Martin was aboard, and after reaching cruising altitude, he turned to me and said, "I understand you are getting ready to retire."

"Yes, sir. I'm retiring at the end of June."

"I'm looking forward to officiating at your ceremony."

I was honored that he offered to conduct my retirement ceremony before I had the chance to ask anyone else. I hadn't told many people that I was going to retire, least of all him. That's the kind of man he was. He took the initiative.

The evening before my official retirement ceremony, Jerri and I hosted a casual buffet dinner with six hundred of our closest friends in the Modern Flight Gallery at the National Museum of the United States Air Force. Dressed in golf attire and a straw hat and holding a glass of scotch, I made a grand entrance in a golf cart with Jerri at my side holding a glass of wine while the band played "Prop Me Up Beside the Jukebox When I Die."

It was a night of fun and entertainment with presentations from my family and units within Air Force Materiel Command. Doug Jankovich collaborated with Jerri and created a DVD slide show of my life and projected it on the big screen for all to see. My family made a scrapbook with pictures documenting my forty-four-year career, including my four years at the Academy. Another memento was a shadow box that contained the Air Force insignia, all my ribbons, medals, and a folded U.S. flag on a backing of Air Force blue. In the middle was a plaque that reads:

Major General Edward J. Mechenbier
For Honorable and Faithful Service to
Your Country and the United States Air Force,
3 June 1964 to 30 June 2004.

I proudly display it above the fireplace in my home.

My official retirement ceremony continued at the museum the following morning, and no, I wasn't hung over, in case you're thinking along those lines. I never drink that much. I've only been drunk twice in my life—once when I was seventeen, and again at age thirty-one in Las Vegas at a River Rats Reunion. Never again!

Perhaps that needs further explanation.

Back in July 1973, the returning POWs were invited to become part of the River Valley Pilot's Association. Until then, membership had been limited to fighter pilots who had flown combat missions over the Red River Valley in North Vietnam. Although they had had several reunions prior to our release, their first really big gathering wasn't until we came home and were invited to join. With a combined attendance of their group and the POW returnees, along with their family members, there must have been more than a thousand people gathered in the Las Vegas Convention Center.

What a party that was. You can bet when that many fighter pilots get together some hell was going to be raised. In keeping with the image of intrepid sky warriors we tried to drink Las Vegas dry, and I did my share by drinking gin. Although I cannot attest to the fact, I was told we hung the girls' bras from the rafters. Jerri told me the next day that I had crawled across Las Vegas Boulevard on my hands and knees while she, Larry Bogemann and his wife, Sandy, stood laughing, probably too drunk themselves to help me to my feet. I have no recollection of impeding traffic other than I found holes in the knees of my tuxedo the next day.

But back to my retirement ceremony.

The folks in the Air Force Materiel Command protocol office, Chief Master Sergeant Mark Plunkett and Martha Smith, worked out the details of my retirement ceremony down to the nanosecond. As you might expect, there was pomp and ceremony with the presentation of the colors and the playing of the national anthem and the Air Force hymn. The ceremony began with the master of ceremony, Colonel Paul Sampson, offering a few brief remarks followed by an introduction of my family, special guests, and dignitaries.

As I walked into the room alongside General Martin I was stunned to see that so many people had come to see me retire. General Martin opened the ceremony with a speech fitting the occasion. He summarized my career from the time I entered the Air Force Academy to my shoot down and spoke of national events that had taken place during that time. Much to my chagrin, he continued the myth by saying, "Despite all the titles Ed has had, I think we'll all agree the one that best describes him is great American hero." I sat there listening while reviewing my notes. "Now, after forty-four years in uniform you will step down from this Air Force. We'll do our best to carry on with the training and mentorship you've established. . . . Strong tail winds, gentle landings forever, God bless you and God bless America."

The audience was asked to stand as he presented me with the Distinguished Service Medal and read the accompanying citation. Then he read the retirement order EX-4852 that booted me out of the service, "Major General Edward J. Mechenbier, Mobilization Assistant to the Commander, Air Force Material Command, Wright-Patterson Air Force Base, Ohio, is relieved from current assignment and is retired effective 30 June 2004, in the grade of Major General by order of the Secretary of the Air Force."

With that it was official. I was no longer a member of one of the greatest organization in the world.

I stood there thinking, this isn't right. I wasn't ready to say goodbye to forty-four years of service. I didn't want any of this to be happening. It's not time yet. I have too much to do. I could have given them a million reasons not to retire me. Didn't they know I was way too important to be kicked to the curb?

Then Colonel Sampson read congratulatory letters from President George W. Bush; Ohio governor Bob Taft; Secretary of Defense Donald Rumsfeld; Congressman Michael R. Turner; and my good friends, Representative Dave Hobson and Senator Mike DeWine.

Jerri was asked to come on stage and was presented a certificate of appreciation along with a bouquet of flowers for her unwavering support and for putting up with me all those years. She gave me a kiss and then saluted me, which brought a laugh from the audience.

Then it was my turn at the lectern.

"When you're getting run out of town on a rail," I began, "get out in front and make it look like a parade. Welcome to my parade.

"My entire career was one of exciting experiences, and I never felt like I was enduring. Even during the time in jail, I was doing my job. The Air Force was thirteen years old, and I was seventeen when I joined, and it's been an exhilarating, evolutionary journey. I've enjoyed every minute of it. I'll miss it. There's no subtle way of saying it. I'm not going to miss the staff meetings and paperwork, but I'll miss the people. That's one of the unique bonds I've been privileged to share all these years."

I thanked my wife and children and gave them an engraved Seiko watch with the Air Force crest and the inscription, "Perhaps we'll have more time to spend together."

Then I quoted the forth verse of the national anthem: "O, thus be it ever when freemen shall stand, / Between their lov'd homes and the war's desolation; / Blest with vict'ry and peace, may the heav'n-rescued land / Praise the Pow'r that hath made and preserv'd us a nation! / Then conquer we must, when our cause is just, / And this be our motto: 'In God is our trust' / And the star-spangled banner in triumph shall wave / O'er the land of the free and the home of the brave!"

I concluded, "Think of those words tonight as you watch the news and know that America and her men and women are still bringing that dream to others. Thank you for being here this morning and letting me play on a team for four decades. Thank you and God bless you all and may He bless you as abundantly as He has me and my family."

With those final words I broke into tears. Yep, like I said, I'm just an old softy.

Next came the furling of my two-star flag. I stood alongside General Martin as my children, Kari and Bo, did the honors. What a proud moment that was to have them in uniform, participating in the ceremony. Standing there watching them perform their duty with such earnestness gave me a sense of legacy. I was inspired to know that my son and daughter wanted to be part of my ceremony.

The ceremony ended with everyone singing and clapping to the Air Force song. That was it. Wow! What else can I say? It's been a great ride. And it only cost me five dollars.

Following the ceremony, 160 of my closest friends and I headed to the Beavercreek Golf Club for an afternoon of golf. The tricksters, Jankovich and Webster, set out once again to make the outing a big event. In league with the superintendant of the golf course, they conspired and had a thirty-foot square blue and white Air Force crest painted on the side of a knoll adjacent to the first hole. Before the game began I was honored with my own mini air show as F-16s from the Ohio Air National Guard and several C-141s from Wright-Patterson gave me a fly-over. I stood with a lump in my throat as I watch the airplane pass overhead, hoping they didn't misfire when lighting their afterburners.

Jankovich had hats made with two stars on the back and tees and coins specially minted for the occasion. He also had custom flags made for each hole. After the tournament he took the number five hole flag, my favorite, asked everyone to sign it, and then had it framed.

It was my day and I should have won. Webster and Jankovich could have arranged it, but the Hanoi Taxi crew won the tournament, and I won the long drive prize of the day. All in all, we had a great time.

For many, retirement means curling up and waiting to die, but not me, I flunked retirement. The next morning, as a retired general officer, I was back at my desk at SAIC with my uniform hanging in the office ready to be pressed into service whenever and wherever needed.

I soon realized I missed being in uniform. Not for the glory of being a two-star general, which certainly isn't a bad thing, but I honestly missed the work, the missions, and the people. On the other hand, I now had only one full-time job instead of juggling two, and no longer had to switch from a civilian suit and tie to a military uniform several times a day. I also had more time to be with my family and to play golf.

Someone once said, "There will come a time when you believe everything is finished, but that is just the beginning." That's why I cherish each life's chapter's ending, because I know there is a new chapter waiting and I look forward to each one. I'm not going to sit in a rocking chair twiddling my thumbs, waiting for the sun to fade over the horizon. I'm anxious to see what the next day has to offer—maybe a hole in one. I can dream with the best of men.

As for retiring—really retiring—I'll work until the day before I die. I think it's important to have a reason to climb out of bed each day—a purpose—someplace to be that gives a person a sense that somebody depends on them. I don't care if it's volunteering time at the local food pantry, being a school bus driver, or working part-time at a major corporation, a person needs to keep busy.

I've seen too many people grow old after retiring. I was saddened to see how old many of my classmates looked when I attended my fiftieth high school class reunion. They acted old, sounded old, and talked old. They talked in the past tense, living through their children and grandchildren's lives. I came away feeling that we had nothing in common.

As long as I can swing a golf club, you'll find me on the links. Golfing is my passion—the perfect game for me. Physically, it helps keep my joints from rusting. Mentally, it demands a level of concentration, and an attitude that keeps my brain cells charged. And golf is relaxing. If I hit a good shot, great. If I not, I don't swear and throw clubs into the lake. The game also provides a spiritual avenue whereby I can escape the daily grind and block out all thoughts and worries running through my head. Walking between shots in the early morning hours, gives me quiet time to say my morning prayers. That's when I have my conversations with God and Kevin.

I still have the hickory shaft golf clubs that Ruth Hinkson gave me as a boy. I'm grateful that in my exuberance to shed my past, as promised to the Weasel Club, that I had enough sense to keep those clubs. To me, they're almost a religious relic and might even be worth something to a collector some day.

Being retired has given me time for leisure reading. I read a lot of management and leadership books. My favorite, one I consider my business bible, is *Good to Great* by Jim Collins about how you can take a good company and make it great. I've read it several times, gleaning something new each time.

What's on my bucket list? Thanks to Al Stafford teaching me to play the guitar in Camp Unity, I came home with aspirations to master the instrument. I bought an Epiphone guitar and continue to learn to play. I'm not too bad, but don't expect a CD with my face on the cover any time soon.

Tribute Flight for the POWs

When the decision was made to have the POW reunion in Dayton in 2006, I thought it would be great, as part of the activities, to have a reenactment of the flight that brought the men home from Vietnam in 1973.

On May 5, 2006, 128 POWs gathered on the tarmac for the reenactment. With Secretary of the Air Force Michael Wynne and Congressman Dave Hobson in attendance, Webster, acting as master of ceremonies, recaptured the moments before takeoff, and as a tribute to the men, said in part:

> The object of this briefing is to allow you to accompany 0177s flight crew on the morning of their historic flight into Gia Lam Airport in Hanoi, North Vietnam, when they brought home some of America's most precious cargo—captured American airman. It was February 12, 1973, and the world's attention was riveted on Hanoi. America's war with Vietnam was over and the long awaited start of Operation Homecoming was about to begin. While the anchormen and camera crews were setting up, the C-141's crew was in the briefing room getting ready to launch.
>
> There will be four airplanes launching today headed for Vietnam. Tails 60177, 50243 and 50236 will fly and land at Gia Lam. The fourth aircraft will orbit on the Vietnamese border as an airborne spare. 0177 will be first to land followed by 243 and 236, which will establish holding patterns and wait for 0177 to land, load, and takeoff. You will have to watch out for anti-aircraft fire, SAMS and interceptors. The AAA is 37mm up to 85mm and the SAMS have a range of up 30 miles. The interceptors should be minimal around your area today. The view of the airport shows that runways 01 and 190 are 6,182 feet long and made of concrete, parking area allows for only one Starlifter at a time due to craters on the taxiways and the ramp.
>
> The major hurdle facing us is the Vietnamese officials declared that there must be a red cross painted on the tail of each aircraft. They figure that American crews cannot respond fast enough. Without sufficient drying time they think the planes will land with smeared red tails. What they forget is we have a special tool called enlisted ingenuity—red speed tape in six feet lengths. It makes a perfect red cross.

Webster ended with words directed at the group:

> Members of the 4th Allied POW Wing, the men and women of the military would like to thank you for your service to your country. Your strength in yourselves and in this country is something that can never be looked down upon. You have lead, trained, and mentored the current generation of military members. You developed and implemented programs that have saved recent captives some of the stress that you underwent and, most of all, their lives.

I was one of the sixty former POWs who lined up for the first flight as we had done in February 1973, waiting for our names to be called before boarding the aircraft. Members of the 445th Airlift Wing escorted each man to the airplane. Once aboard we took a seat and waited for the engines to power up.

As the Hanoi Taxi lifted off, one of the flight crew called "Feet wet" just as was done thirty-six years earlier. We cheered and hugged each other as a flood of memories carried us back to that fateful day. Kevin was there, of course. Having kept in contact with him throughout the years, it wasn't as though we needed to catch up on the past thirty-six years, but we went along with the camaraderie.

When the emotions settled, some sat in silence with their own private thoughts, while others sat retelling stories. Some chose to walk around looking at the many photographs and memorabilia attached to the cabin walls. Vietnamese and American flags were displayed along with a plaque with the names of the original crewmembers put there by Master Sergeant Henry Harlow who had become the historian and keeper of the flame for 66-0177.

Lieutenant Colonel Tim Baldwin piloted the first plane, (Lieutenant Colonel Steve Johnson flew the second flight) and I had the honor of sitting on the flight deck for both flights. We flew around the Dayton area for about an hour, and since many of the men had never been on the flight deck of a C-141, and certainly didn't have the opportunity on the flight out of Vietnam, I acted as a tour guide to the steady stream who came up to take a look at the instrument panel and to ask questions and talk with the secretary of the Air Force.

Following the flight we landed to cheers and "Welcome Home" banners as each man was escorted off the airplane into the welcoming arms of waiting wives, children, and friends. There wasn't a dry eye anywhere.

Saying Goodbye to the Grand Ole Lady

Although the Lockheed C-141 Starlifter, tail #66-0177, was only one of ten aircraft employed to fly POWs out of Vietnam in 1973, being first to land at Gia Lam and first to lift off with it's special cargo, made this particular aircraft the iconic symbol of freedom for POWs and deserving of a place in history. So, on May 6, 2006, the Hanoi Taxi was retired with honors at the National Museum of the United States Air Force in Dayton, Ohio.

Following the POW Tribute Flight, the next morning the C-141 sat waiting on the ramp in front of the 445th Airlift Wing Headquarters. The sun was shining at 0900 when I arrived on the flight line to be part of the ceremony.

Major Steve Schnell sat in the right seat with Colonel Steve Johnson in the left seat as pilot. I was invited along as a guest because of my long history with the Hanoi Taxi. The flight included Lieutenant General John A. Bradley, commander of the Air Force Reserve Command; and General Duncan J. McNabb, commander of the Air Mobility Command.

Also aboard was Team 177, the service members responsible for finding the history of the airplane and turning her into a flying museum: Senior Master

Sergeant William H. Warner; Master Sergeants Glen D. Janus, Robert A. Brown, Stephen C. Armstrong, Henry Harlow, David Dillon, and Jeff C. Wittman; and Technical Sergeants David E. White and Carl D. Hayden.

I sat in the back knowing that the pilot and copilot were going through the avionics preflight as I had done so often before and knew that it would be the last time that check would ever be made on this aircraft. It would be difficult, if not impossible to know how many times that yellow checklist had been used during the airplane's forty-three-year history.

Two base fire trucks formed a line and made a traditional water arch as we taxied to the runway. Schnell pushed the controls forward and the airplane lifted off at 1230.

With approximately forty thousand pounds of fuel on board, we flew a pattern around the air base for about a half hour to burn off the excess fuel and to give the hundreds of people waiting at the museum for the formal retirement ceremony a chance to see the Grand Old Lady in the air one final time.

I was looking out the window, remembering all the times I had flown this aircraft, when General McNabb came up to me and said, "Why don't you go on the deck and fly her in?"

"I'm a civilian now and just a passenger."

"Well, you're back on active duty today. Go fly the plane."

Never one to argue with a four-star, I did what I was told, delighted to have the opportunity to fly that beautiful bird one last time. It had been more than two years since I had flown her and it felt good sitting in the left seat again. Knowing that I was part of history, the last POW flying the Hanoi Taxi on her last flight, I couldn't help but become nostalgic. After all, we had a long history together, and I was aware of the honor McNabb had given me, and I was grateful.

After thirty minutes at the controls, I relinquished control of the airplane to Johnson who made an exceptionally smooth landing in spite of the 10-knot winds. He taxied the airplane to the reviewing stand where the crowd waited anxiously for the official retirement ceremony to begin.

As the crew went through the final engine shutdown checklist, I listening as the hum of the engine went silent forever. The C-141, tail number 66-0177, that had served with honor since her maiden flight in the 1960s, had played as an important part in our country's military as any airman, foot solider, or seaman. She had earned her place in history. Now she would fly no more.

As the band played patriotic music to a teary-eyed crowd, a well-timed flight of F-16s flew over in the missing man formation, adding a poignant touch. As part of the dedication, General McNabb paid tribute to her with these words, "It was a magnificent aircraft. 0177 will carry the memories of all who flew on her. Rest well. You have been blocked into history, 100 percent mission complete."

Ross Reynolds, a representative for Lockheed Martin, said in part,

> One of the proudest moments in the history of the great C-141 in 1973 was when we were able to bring back the POW from Hanoi to Clark Air Force Base. . . . I would say that it is absolutely fitting that this aircraft, this tail number, take its last victory lap. . . . The C-141 has been the backbone of this country's Strategic Airlift Force for more than four decades. On behalf of the men and women of Lockheed Martin yesterday and today and to everyone who has been associated with the aircraft, to the Hanoi Taxi, in particular and the C-141 Starlifter in general, we say with the utmost respect and admiration, well done, good and faithful service.

I'm sure that for many, watching the C-141 make her last flight was a highly emotional experience. That was especially true for many men who had flown her and the crewmembers who maintained her, but for Master Sergeant Henry Harlow, part of Team 177 it was much more—it was an end of a love affair. Henry had devoted countless hours researching and preserving her history and had developed an emotional attachment to the Grand Old Lady. To him her retirement was personal.

As everyone was milling around, talking and enjoying the celebration dinner that evening, I watched Henry as he respectfully, almost reverently, walked to the airplane, put his hand on the fuselage, and bowed his head, saying goodbye to his Lady.

CHAPTER 24

Family and Friends

JERRI HAS ALWAYS BEEN MY ROCK. She stuck with me through all those black years, praying for my safe return. She has put up with my craziness, my love of flying and golf, and my days and weeks away from home on deployments and while traveling for work. As I said in my retirement speech, probably the reason we lasted with each other so long was because I was gone half the time.

Jerri and I finally got to celebrate our first anniversary together in June 1973. Our first anniversary in 1965 I was at the New York World's Fair flying a T-38 Talon on a cross-country training flight. Our second anniversary, I was in Libya upgrading to an aircraft commander while Jerri was in England. The next six anniversaries were spent locked in a Vietnamese prison.

Our lives at that point were separate but parallel. Fortunately, Jerri has an independent nature and with me being gone so much, was quite capable of managing our finances. When I went to Vietnam, she took the checkbook and I haven't seen it since. That's works well for us.

Whatever the reason, Jerri and I have a strong bond. Of the 591 POWs that returned home after their ordeal almost half are divorced. Some men, even after thirty-five years, find they still can't bridge that gap and allow even their wives to penetrate the barrier they built around themselves. I know Jerri feels she has never been let inside that protective circle. It's not that I push her out, it's that I don't see a reason to bring her in. Emotionally, there are parts of me that no one will ever see. For that matter, there are still parts of me that even to this day I don't see unless a sight, sound, or smell triggers memories and emotions from that time in my life.

Raising four children didn't leave Jerri much time to be part of my career, but she tells me she wouldn't have changed a thing. She's the glue that holds the family together. We have grown as husband and wife, as most all couples do, from a young couple crazy in love to two friends who have settled into a comfortable relationship. I can remember as a kid sitting around the radio listening to Norman Vincent Peal and hearing him say that you know you have arrived in a relationship when you truly like your partner. Jerri and I are beyond the "I love you" stage in our marriage. I can say without hesitation that we truly *like* each other. I guess that means overlooking each other's habits and faults and still wanting to be with them above anyone else.

I couldn't be more proud of my children. I wish I had had more time to be with them as they were growing up, but holding down two full-time jobs was like juggling hot potatoes, as I tried to maintain a balance between job and family and being with them for their activities. Thanks to Jerri's nurturing, they have all become fine adults. They are successful, each doing something different. Mahli is a professor at Kent State, teaching technical writing and English. Kari and Bo followed in my footsteps and joined the Air Force; and both married members of the Air Force. Tai is a superb mom and homemaker, married to a man who works with computers. There are grandchildren and I'm certain more will come along through the years, adding to the Mechenbier clan.

I'm not sure of the contribution I have made in our children's lives, but I'd like to think that I've been a friend as well as a good father. I must have done something right because I noticed that as they grew, at some point I stopped talking to them as children and started talking to them as adults.

In keeping with my father's tradition, I had our children line up on the stairs and made them wait until Jerri or I took their photograph as they came down the stairs to open their Christmas presents. I hope they had as much fun as I did anticipating what Santa had brought. I make their Christmas cards every year, and this past year I wrote inside them, "The colors on this card are red and green. The red is for the love in our hearts for you always and the green is in your account at the bank."

My mother died in 1982 from lung cancer at age sixty-two. Mom played a big role in my development by fostering a sense of dedication to family. She kept the household running smoothly when my dad was working three jobs. Even though she wore many hats and stayed busy, she was always there when we needed her. All of us knew that if anything bad happened we could go to her and she would "fix" the problem.

My father died at age seventy-six in 1993 of a heart attack. He was a fine man and dedicated to his wife and children. I am thankful that he was always there for us. He showed us the importance of family and instilled the moral values, work ethic, and social standards that I live by today. I cherished his love and the time he spent with us as kids raising us to be solid, law-abiding citizens with strong religious values. Through his example, I learned that family is the greatest asset a man can have; enjoy them, always make time for them, and cherish that time together.

When I went to see my father after his second quadruple bypass surgery, Dr. Karl Borsody, the famous heart doctor who lived in Dayton and who performed the surgery, told me to tell my father that I loved him. I told Dr. Borsody that my father knew I loved him. He said, "No, tell him that you love him."

I had always acted the part but it wasn't until that moment that I came to fully appreciate what my father had done for me and my sisters and brothers. Remembering my promises as a member of the Weasel Club and what some of the other guys had said about their relationship with their fathers, I had no trouble saying, "I love you."

Our Vietnamese Friends

In 1976 Jerri saw an advertisement in the newspaper inviting the public to a local church for dinner being prepared by a South Vietnamese couple that had just come to this country. Jerri said, "I bet you've never had good Vietnamese food." She was correct on that point. Boiled pumpkin with bits of unidentifiable floatables didn't qualify as real cuisine. "Sounds good," I said. "Let's go."

At the church we met Oanh and Hiep Pham. Their limited ability to speak English and ours to understand Vietnamese made for strained communication, but we managed and discovered that they had aspirations of owning a catering business. Jerri and I were taken by their kindness and gentle nature, and in the years to come our friendship grew, as did their business. In fact, their restaurant became so successful that with our help as silent financial partners, they were able to sell their first restaurant and open a second larger one, and with that success they sold it and bought a third. Our participation in the business wasn't just silent; every Friday and Saturday night for three years I washed dishes—old habits die hard—while Jerri acted as waitress and hostess.

In additional to our relationship with Oahn and Hiep, Jerri thought that it would be nice for Mahli to learn about her heritage and hoped that when she turned sixteen she might like to work for Oanh.

Wrong. At sixteen, she went to work at a pizza place.

Eating soup with chopsticks can be a messy affair, but it gave me the opportunity to be creative using my artistic skills to make bibs for the first twenty-four patrons, the charter members of what we called the Bomb Tie Club. Using puffy paint, I made bibs and personalized them with the customer's names and drew a caricature of them that represented their individual personality or characteristics. What fun that was! Everyone loved them. Unfortunately, when Oanh and Hiep sold the restaurant, the new owners insisted on keeping the bibs.

The restaurant acquired some unsolicited publicity when Robin Leach, from the television show *Lifestyles of the Rich and Famous*, came into town for a presentation to the Dayton Women's League. The night before, I took him to the restaurant for a meal and introduced him to Oanh and Hiep. The next day, Robin announced that he had gone to "the world's most wonderful" restaurant and suggested to the audience they go to General Mechenbier's Vietnamese restaurant.

In 1980 Oanh mention that she could buy several hundred acres of land on Indian Ripple Road for sixty-four thousand dollars and asked if Jerri and I would like to invest thirty-two thousand dollars in this real estate venture. We didn't have that kind of money and had to turn her down. She bought the property that was later developed into a huge shopping mall called the Greene, which turned out to be worth far more than the original purchase price. Oh, what could have been if only we had a crystal ball.

Over the years we continued to maintain a business and social relationship. Oanh and Hiep often asked us to go to Vietnam with them, but each time we declined. With four small children and Jerri's mother living in a nursing home, there was no way we could take such a long trip.

Return to Vietnam With Jerri

Several years later, Jerri surprised me with the news that we were going to Vietnam. She and Oanh had formed a conspiracy and had bought tickets and applied for passports and visas. Although I had been back to Vietnam in 2004 on an official military mission to return the two bodies of our service members, I had no desire to return as a tourist, and told her so. Jerri insisted that we should go since we had nothing holding us back; our children were grown, and there was no good reason not to go. She said it would be fun to look for the jail where I was held near the Chinese border. When I hesitated she sweetened the deal with a promise that I could play golf while there. That settled it. We packed and headed for the airport.

As soon as we landed in Saigon the local authorities confiscated our passports. Anh Sau, one of Hiep's cousins and our host, explained that there was an official in charge of the neighborhood who needed to know who we were and what we were doing there. Typical Communist.

The flight had taken almost fifteen hours and we were exhausted by the time we arrived at our hotel. Jerri and I fell into bed around midnight only to be awakened by trumpets and drums blasting "Bésame Mucho" in the street below our window as mourners took part in a Vietnamese funeral ritual. Because they worked all day they observed funerals at night. The music went on for three nights until four o'clock each morning while they paraded up and down the street.

As we drove through Saigon, I noticed much had changed since 1973, and much had stayed the same. Kids talked on cells phones in front of Internet cafes while little old women swept the sidewalk with Halloween like brooms. Women carried buckets on long poles balanced on their shoulders and worked in rice paddies alongside of water buffalo. Entire families rode on motorbikes with babies perched between parents, and Western commercial advertisements decorated the sides of buildings.

With the prospect of buying land that was soon to be available for development, we took a ferryboat trip across the Saigon River to a peninsula where Oanh and Hiep owned property. Land that had sold for around thirty thousand dollars for a two- to three-acre lot, doubled in price overnight once word got around that two white people were interested in property there.

Leaving Saigon, we flew north to Hanoi, landed at Noi Bai Airport, and was met by Do Van, who served as our host. Among his family he was known as Number Two son and lived in England and worked for the BBC (British Broadcasting Company). (All the children are called by their birth ranking.)

I quickly became aware of the stark contrast between Hanoi and Saigon. Hanoi was much cleaner. French architecture still influenced the city and brightly painted multilevel houses had porch-like patios on the top floor with the open lower floor used as a garage or as a place to sell their wares.

While in Hanoi we stayed at the Zenith Hotel which, by American standards, would rate as a three-star hotel. We had a nice room with a television. Yes, they finally had television, and with BBC and CNN beamed by satellite from China, Singapore, and Australia, we had many programs from which to choose.

Do Doan Chau, Number Six son, showed us around Hanoi and took us to see where the Hanoi Hilton once stood. Portions of the Hoa Lo prison had been torn down and a twenty-five-story office tower built in its place. How ironic to discover that the very place where my prison cell once stood now included a golf shop. How bizarre is that?

Jerri did a lot of shopping—a lot of shopping. She fell in love with a painting by Do Doan Chau, and we each bought one of his works. We also purchased a piece of lacquer art created by an internationally acclaimed artist named Bui Tuan Thanh, a friend of Chau's.

We trekked through old Hanoi, with its quaint shops where people still cooked on the sidewalks. On Sunday, after attending Mass at the Cathedral, Jerri and I roamed around looking for one of the old prison camps. We found the railroad tracks where Gold Tooth Fairy put Kevin and me on display for the train passengers. We followed the tracks until we came to a high wall with glass shards embedded at the top. I figured it had to be the Plantation.

It was nice being on the outside of that wall.

The Paul Doumer Bridge still stood despite our efforts to destroy it. We located the underground factory, still as dingy as I remembered it, if not more so. The only emotion I had was disappointment. The buildings had not been preserved. They didn't even think enough of me to place a plaque on the door saying, "Ed Mechenbier slept here."

On our third evening in Hanoi, Jerri and I were invited to dinner with Oanh's large family, including Number Five son, Hai, the minister of Culture, and Do Doan Phu, Number One son, a former North Vietnamese army colonel who had been in charge of logistics and construction at the Vu Chu railroad yard in Kep where I was shot down. How very odd to be sitting with the man who was the commander of the railroad yard I had bombed.

As a retired colonel, his house and food were provided for, and he was given a driver to take him anywhere he wished to go. Apparently being a retired colonel in Vietnam made him a minor god.

As a gesture of friendship, Number One son stood and saluted me, and then

Number Five son said, "You fly the airplanes and we ducked."

"We ducked too!" I replied with a smile, letting him know that we were also being shot at.

During dinner, Jerri happened to mention that I was number one son also, being the oldest son in my family of eight.

"You are a greater Number One son because you are number one of nine and I am the Number One of eight," I said, and then holding my wine glass, I added, "As one oldest child to another, I toast you as being the greater number one." Then he added, "The colonel salutes the general."

There we were, once bitter enemies, now sitting across the dinner table, face to face, toasting one another.

The conversation turned to the bombing and how the war had transformed their country. He explained that their feelings toward us were never personal and they never held our actions against us. They always looked at the war as something precipitated by the American government, not the men who fought it. In fact, everywhere we went during our travels, Henry always introduced me as his excellency, the air pirate. It got to be a joke. I felt self-conscious at first and asked him why he felt he had to tell everyone that I was a blackened criminal who had bombed their country. He smiled, having a grand time with the situation, so I put my hands together as if in prayer and bowed as is the custom.

Thinking it would be a magnanimous gesture for two old enemies to meet in friendship, Henry had made arrangements for me to meet Rabbit, not knowing to what degree the Rabbit had been my tormentor. This was the only time I had misgivings about seeing the people who had held me captive. Unfortunately, Rabbit died two weeks before the meeting, so his son was asked to attend instead. I was taken back by his resemblance to his father, but there were no feelings of hatred or animosity or earth shaking flashbacks to that dark time in my life. There was only a flicker of memory of the more unpleasant conversations I had had with the former camp commander. His son and I had lunch and a few drinks and then parted.

The final leg of our trip took us back to Saigon where we spent several days sightseeing and visiting the Crystal Palace where Oanh once lived. We took a day trip to the Vung Tau Beach Resort where we stayed at a five-star hotel that cost only sixty-five dollars a night. The food was fantastic, but unsure of drinking the water, Jerri insisted we drink only beer with meals. She got no argument from me.

We had been in Vietnam fifteen days and I still hadn't seen a golf course and was beginning to have golf withdrawals. Jerri didn't protest when I suggested it was time to leave, and with our business complete, we said our goodbyes and bid farewell to Vietnam.

The Pacific Return Golf Tournament

The eight of us who played golf on the trip back to Vietnam get together every year to continue the Pacific Return Golf Tournament. We meet at the Oglebay

Golf Course in Wheeling, West Virginia, where Ish continues keeping score to suit his purpose as names are added to the trophy.

Every year someone has a small souvenir made for the event. One year I had shirts made with Pacific Return Tour embroidered on the sleeve, another time I had shot glasses made from 30mm shells, and another time it was 40mm shell cut down and polished and made into ball markers.

As a more permanent memento, Doug Jankovich and John Cherry designed a beautiful wooden chest that rotates every year, according to who won the last tournament. Carved on the top of the lid is the patch, also designed by John, that symbolizes my Vietnam experience—a map of Vietnam with a star over Hanoi, an F-4 flying into Vietnam, a C-141 flying out of Vietnam, and the POW/MIA flag all inside a circle with the dates of my shoot down, release, and the date of my return flight. On the inside of the lid is a plaque listing the names of the eight Freedom #159 Crew members/golfers who flew the repatriation mission. The chest is lined with crushed blue velvet and in the center sits a bottle of Red Baron Flugbegleiter schnapps surrounded by eight small pewter cups each engraved with our names.

At the end of each tournament we raise our cups in a toast.

> To the crew, may next year find us all better, wiser and younger, at least mentally.
>
> To the NASA computer whose internal workings have again befuddled us all to the point of not even caring who won the game.
>
> To those who have gone before and to those to whom God did not give a Homecoming and to those who have departed since last year.
>
> And finally, to the numerous sips, gulps, quaffs and guzzles to finish the contents and move into the bar for some good drinking.

Our last trip to West Virginia in 2011 ended with me in the hospital staring down death's hallway. The eight of us had made a stop at Salt Fork in Cambridge, Ohio, for a round before heading on to West Virginia. Although I felt lightheaded from time to time, I finished the eighteen holes. Afterward we drove to TJ's Sports Bar in Wheeling for dinner and drinks before driving up the hill to the Oglebay Golf Course.

While the guys went to the bar, I returned to my cabin and went to bed. Some might say that my errant ways had caught up with me, but I had been good and hadn't been drinking all that much, so I was dismayed when I got up the next morning feeling like crap. I thought, perhaps, I had eaten something that didn't agreed with me and told the fellows to go ahead, I'd catch up with them for the afternoon round, and went back to bed. As the morning wore on, however, I felt worse and began to pass blood.

When John Cherry texted to ask how I was doing, I told him I still didn't feel well enough to play the afternoon round. Instantly he knew something was wrong, and in no time was at my door. I told him I thought I had a problem and needed go to the hospital.

We were loading my clubs and suitcase into my truck when I passed out. John carried me into the clubhouse and got me a glass of water. We discussed going to a local hospital, but not knowing what kind of care I would receive there without my medical records, I told him I felt well enough for him to take me back to Wright-Patterson.

As he poured me into my Toyota I passed out again. With that he took off as if he were flying an F-4. He literally flew the road as I drifted in and out of consciousness. We left Wheeling around one o'clock and I was in the emergency room at Wright-Patterson Medical Center at four, a distance of about two hundred miles. At some point Cherry called Jerri, but I have no idea when she arrived at the hospital because by the time we got there, I was completely out of it.

I won't go into the details, only to say they found I had a bleeding ulcer and had lost a lot of blood. Although the hospital staff was top notch, there was one small glitch. I wanted to to be completely asleep. I have a strong gag reflex and knew I wouldn't behave if only semiunconscious, but they gave me just enough knock 'em out juice to put me into a semiconscious state, and when Dr. Gregory Mies began to insert the scope down my throat, in my half-drugged condition, I saw a man standing before me wearing a porcelain armored breastplate and a hockey mask. I began thrashing around, yelling, and verbally assaulting him, calling him dumbsmack and dipscrew. Luckily, he was an Academy graduate and recognized my rants. After the doctor suffered "contusions and lacerations" from not listening to me, they knocked me out.

Since I couldn't complete the golf tournament, and because no one had won the Mechenbier Trophy, the guys decided the tournament would be replayed at a later date. Once again proving the extent to which those guys will go for each other. Nice bunch.

The Mechenbier Hope Diamond

It wasn't until I returned to the Air Force Academy for our tenth reunion that I was told the full story of what happened to all my possessions after I was shot down and ended up as a guest of the North Vietnamese.

When someone was listed as MIA or KIA, a summary court's officer, acting as a legal representative with the fiduciary duty to tie up loose ends, is appointed to collect the man's belongings and ship them home to his family. I had among my belongings a black star sapphire ring that I had bought for Jerri while on R&R in Thailand. Before the first summary court's officer could fulfill his duty, he was shot down and killed. Another pilot was assigned the duty and he, too, was shot down. You can imagine the long line of men eager to raise their hand and volunteer to collect my belongings and send them home.

Because the black sapphire ring was among my possessions, it became known as the Mechenbier Hope Sapphire, a play on the Hope Diamond, and a blackened symbol of death.

First Lieutenant Don Kammerer was next in line for the ill-fated task.

While going through my footlocker, he noticed that the ring was missing. He took up a collection within the squadron and bought another ring and sent it, along with my things, to Jerri. She still has the ring tucked away among her treasures and I'm happy to say Kammerer lived through the ordeal.

Speaking Engagements

For many, speaking in public is a frightening experience. For me, public speaking is an opportunity to educate people about today's Air Force.

Although I consider myself a shy person, I'm not shy on stage. People find me entertaining and more importantly, I'm cheap. I don't require remuneration for my speeches, but if a round of golf is offered, I never turn it down.

During the early years, people wanted to hear what it was like to be a POW—to drink in all the gory details about the torture and how I survived those long years in captivity. They wanted to see and feel what it was really like. Knowing there was no way I could talk about that time in my life, I tried to discourage questions about being held prisoner. I did, however, feel an obligation to tell them what it was like being held captive under a communist regime, not to get sympathy, but as a means to enlighten them about the strength and resiliency of the human mind, body, and spirit. I kept my talks light and told them about the humorous side of living with a bunch of crazy pilots. With my personality, humor comes easy.

Over the years I have given hundreds of speeches to thousands of people. I get invitations to speak at Veterans Day and Memorial Day ceremonies, military retirements and promotion ceremonies. I have been the Master of Ceremony and Keynote Speaker for many professional and business organizations and service clubs. I especially like talking about today's Air Force and the opportunities military service offers and about the pride of wearing the uniform of the United States Air Force.

The presentations I enjoy most are those I give to military organizations such as the Squadron Officers College, the Air Command and Staff College, the Airman Leadership School, and of course, the Air Force Academy. In 1974 I was asked to be the guest speaker at the Air Force Academy's Homecoming, which was also the first time all eleven 1964 graduates who were POWs attended.

When talking to military groups, I stress the importance of the military's legacy and heritage. My main topic is leadership, responsibility and organization, and adhering to the Core Values: Integrity First, Service Before Self, and Excellence in All We Do. That's where I'm in my element; I'm talking to my "Peeps."

I am still active in the Association of Graduates of the United States Air Force Academy and am often a guest speaker there. I tell them, no matter who they are, no matter what they have accomplished or want to accomplish, they are what they are because of the Air Force Academy. Whether they've become the president of General Motors or a four-star Air Force general, it's because of the values they adopted and the lifestyle they were taught while at the Air Force Academy. Be grateful for it. And, oh, while you're at it, say thank you with a financial contribution.

When speaking to the Cadet Wing, I tell them about my Academy training as having been instrumental in the way I, and other graduates, conducted ourselves while being held captive in Vietnam. The skills we developed, the emotional maturity that came from being a cadet and the values that were instilled in us served us well. Those skills still serve me today.

I can look at the ten Academy classmates who were imprisoned with me and know they were all rock solid. Even though we were young lieutenants, we provided a role model to others. We had discipline and conducted ourselves with integrity, never dishonoring our country. Our training reflected credit on the Academy and the POWs as a group.

In 2000 I was invited to speak about my POW experience to a local high school's first-semester Spanish class. I started with the sentence I memorized from Dave Luna in class back at Camp Unity. "*Cuando yo vuelvo a mi casa en los Estados Unidos de América, voy a poner mi hueso del amor entre los piernas de mi esposa.*" I said the sentence very fast so that the students couldn't grasp it, but I knew the teacher caught it when I heard her gasp.

I followed that with, "I know I said that faster than most of you can understand, so ask your teacher later what I said."

I don't know why she never asked me back.

As I said, I don't get paid to speak, but once in a while I accept a stipend, which for a speaking engagement at Northwestern University School of Communication in Chicago was thirty-five hundred dollars. Jerri and Mahli thought it would be a great idea to accompany me on that trip and go shopping. I was delayed leaving for Chicago because of another commitment, and Jerri and Mahli went on ahead. When I arrived at the hotel, Mahli met me at the door with bags and bags of shopping spree bargains. They had spent every cent of my speaking compensation.

On another occasion, I was lecturing to colonels and brigadier generals select at the War College at Maxwell Air Force Base in Alabama when an officer came up to me, introduced himself as Colonel Jeff Frye, a wing commander, and said, "General, I would love to have you come to my base and speak to my wing and squadron commanders."

Of course, I said yes and told him to give me a call any time.

A few months later Colonel Frye called to ask when I could come speak to his airmen. We chatted for a while, making arrangements for my presentation, including a round of golf. "Well, Jeff, I'll be happy to speak to your commanders, but one question, where is your base located?" I asked.

"We're at Ellsworth Air Force Base in Rapid City, South Dakota."

I had just agreed to go to South Dakota in February, and it just so happened the day after I arrived, a blizzard socked in the base. There would be no round of golf. I did, however, see the golf course as we drove past.

During a speaking engagement in Ottawa, Canada, in 1990, I was asked to conduct a half-day seminar to the Chaplains in Combat on their role supporting the spiritual needs of troops in combat.

I was met at the airport and whisked to what I was told was the most expensive restaurant in the city. Knowing they were going to take a huge cut in their budget for the coming year, and not wanting to leave any money on the books that would justify such a cut, they wanted to blow all their funds for the fiscal year ending in February. So the thirteen of us, including the Catholic and Protestant chaplains, their senior staff and I dined on champagne, wine, lobster, and steak.

I gave my presentation the next day. When the budget came out a few days later, they got a 15 percent increase. When I returned the next year they were all too happy to wine and dine me again. They gave me credit for their enhanced budget because of my sterling presentation and presented me with a plaque that said I had ecclesiastical powers in all providences of Canada to any one willing to honor them.

My ecclesiastical powers followed me back to the states when the facility manager at SAIC, Nancy Massey's husband, Jim, died in 2003, and Nancy asked if I would say a few words at her husband's funeral. To refresh my memory, she gave me a couple note cards with information about the children and her husband. I was to talk about what a wonderful man and great provider Jim had been.

When Jerri and I arrived at the funeral home in New Carlisle the funeral director approached. "Reverend, are you ready to begin the service now?" he asked.

I explained that I was a friend and had been asked to just say a few words. He told me that Nancy had identified me as the reverend that was to officiate. What was I to do? I gave a thirty-minute impromptu funeral service.

After the service Jerri and I arrived at the burial site and took our time getting out of the car, trying to fade into the background. Jerri began laughing when she saw the funeral director standing under the tent motioning for me to come forward.

"Are you ready to conduct the gravesite service?" he asked.

I ended up doing the entire funeral ceremony for Nancy's husband.

A couple years later when Nancy died, her family called me and said that since I had done such a fine job at their father's funeral would I please conduct the service for their mother. I ended up doing her service, too.

If anyone should ask, I'm in retirement now as an Apostle.

In 2008 I was asked to speak at the 560th Charging Cheetahs' Thirty-fifth Reunion Dinner. Having had presidents, generals, corporate leaders such as Ross Perot, and other big shots as their guest speaker, I felt greatly honored by the invitation.

As you might imagine there are a lot of ceremonies, customs, and traditions in the military. While a guest speaker at the Air Base Wing Dining In at Edwards Air Force Base, Lieutenant General Richard Reynolds and I had the pleasure of hosting a visiting exchange officer from Bolivia who had never experienced our ritual of the Grog Bowl.

The history of the Grog Bowl is not clear. One opinion is that it began when Cavalry Troops shared their spirits; another is it began in England. Who knows—but the tradition of meeting socially in a military setting in a spirit of camaraderie caught on and has become a main feature at the Air Force Dining In.

Always happy to share our traditions, I led the procession to the Grog Bowl, filled the glasses and gave the salute. "A toast to those who fly our skies, we hope we taught you well, carry on, fly safe and we envy you to hell."

I watched our Bolivian guest as he drank the contents of his glass without it leaving his lips, and then in a fashion to prove he'd drained the glass, tipped his glass upside down over his head, and stood there with a huge, silly grin pasted on his face. I still have the photograph to prove it.

CHAPTER 25

On a More Personal Note

The Role Religion Plays in My Life

I HAVE ALWAYS HAD A DEEP FAITH IN GOD, and religion is still central to my life. Raised in the Catholic faith, I attended Sunday Mass with my family and sat proudly listening to my dad sing in the choir. After church Mom always prepared a delicious dinner. Attendance at the table was mandatory as we gathered to say grace before the meal and knelt to recite the rosary afterward.

I was an altar boy from the time I was a fourth grader in Albuquerque and still help serve Mass today. I say the rosary daily, attend Mass faithfully on weekends and even manage to go to church a couple of times during the week. Am I too holy or too good compared to those around me? Absolutely not; it's just I have my faults and need the extra help.

For me, my faith is a personal matter. I don't wear my religion on my sleeve. I may not be the model Christian, but I have tried to live an exemplary life according to the teaching of Christ and the Church and use the foundation of my religion in a lot of what I do. If nothing else, I keep my religion in the forefront of my mind to tell myself that if I failed in doing something today, or I didn't do it quite right, I will try harder tomorrow.

I believe everything happens for a reason. Having come within seconds of dying in my burning aircraft, I still pinch myself every morning, happy to be alive, and wonder why was I spared. Why did I survive being held captive when others didn't make it home?

There's a reason I didn't stay with that airplane a second and a half longer and ride it into the ground, and why I wasn't beaten to death by a guard. I have always taken the attitude that God got me through that ordeal for some special purpose, and I'm honest enough to believe I didn't do it on my own. I am always thankful the good Lord gave me the privilege of a Homecoming that was denied so many gallant men.

Disappointments

Frank Sinatra may have had regrets, as he claims in his song, but I never have. I do have disappointments, however.

One of my biggest is discovering that personal values have become shallow. Commitment to others is fleeting, and there is too much looking inward, too much self-gratification and greed; too much, "everything for me." It seems people have their hand out, palm up wanting something, rather than stretched out to others in a gesture of giving. For better or worse, the accumulation of material possessions seems to many people more important than spiritual fulfillment. Family values have also eroded. The number of divorces, leaving children without the stable foundation of a mother and father in the home is disturbing.

There also seems to be a lack of integrity. Lying to save face has become the norm, and strong moral principles have gone by the wayside.

It bothers me to see a man wear a hat inside a restaurant, and I don't like seeing men not opening doors for women. This is not a sexist attitude, but a common courtesy.

I am disheartened to find that the discipline, dedication, and the commitment to the military has deteriorated. The respect for leadership and our heritage, the sense of esprit de corps, and pride in wearing the uniform has given way to an "I don't care" attitude. I have seen young airmen in uniform outside without wearing their hat; their shoes unpolished and uniforms in need of a good pressing. I've heard airman say "Huh?" to an officer, not "Sir." It's sad.

I personally take exception to the idea that cadets at the Air Force Academy are called freshmen, sophomores, juniors, and seniors. The Academy is not college. Cadets hold a military rank, receive pay, and come under the jurisdiction of the Uniform Code of Military Justice. That said, the Air Force Academy is still the best academic institution in the United States. I have never regretted the decision I made, or looked back, or had second thoughts about entering the Academy. It is because of the Academy that I have been privileged to wear the uniform of the United States Air Force for forty-four years. It was an unbelievable learning and growth experience. I cherish those years, the people, and what I learned there. Wow, how can I not be grateful? I'm proud to be a "Ring Knocker."

I have never visited the Vietnam Veterans Memorial wall in Washington, and won't. I don't need to be reminded of the war. I admire the sacrifice made by those men and women whose names are engraved on that wall. I honor them and am grateful for their service, but I don't think that mere names on a semi-buried wall is a tribute that adequately reflects the contribution of those who served. The memorial is a dedication to those who gave their lives, and deservedly so, but until 1984 when the Three Soldiers was commissioned, there was nothing that saluted the men and women who proudly served only to come home and suffer in VA hospitals or on street corners today.

My focus is on making sure that Vietnam veterans who never got the recognition they deserve are valued and appreciated for their service and sacrifice. Although there has been an effort to heal their wounds, there will always be scars. And what is being done today to say thank you to all those who served is too little too late.

My Physical Condition Today

If I were to describe my physical condition today I'd have to say it's excellent going on superb, if you don't count the arthritis in both feet, left wrist, upper and lower back; the degeneration in my knees and shoulders; the scar tissue between my lungs and esophagus, and the continued pain from a spinal compression.

I sit and I stand more slowly, and walk with a slight glitch in my step, but I'm doing pretty well, all things considered. I keep physically fit because I know it's important. I don't allow these minor inconveniences to hinder me from living life to its fullest. I stay active because I don't want to end up with some of the problems other POWs have had. I'm determined to wear out before I rust out, and despite what has happened to my body, I still do pretty darn good.

The doctors tell me there's no guarantee that today's medicine will make the pain in my joints any better. They tell me that I can wait a year or two until I can't walk, when medicine is more advanced and less invasive to have a better chance of success at reducing my discomfort. That's plain crazy.

I have nightmares from time to time, not about being tortured, but about not being where the Vietnamese guards expect me to be. I dream that I'm at a ball game in the United States and find myself confronted with having to get back to my cell before the guards miss me.

As a fighter pilot, I always prided myself on having lightning fast reflexes, able to react to any situation almost before it happened. It wasn't until I decided to go back to school and take a computer programming class at Wright State University, that I noticed my response time had slowed. At that time data processing was still done with computer cards stacked in trays. While at the computer center one evening, I was taking my tray of cards to the sorter and bumped into another student and dropped the cards. Thirty years earlier I would have caught the cards in midair and had them back in the tray in the correct order in the blink of an eye, but the cards hit the floor before I could react. All I could do was look at the scattered cards and let them lay. Like Humpty Dumpty, there was no way humanly possible to put those cards back in order again.

Each year I still travel to Pensacola, Florida, for my annual physical at the Naval Operational Medicine Institute. As part of a medical study group, I spend several hours talking with doctors as they poke and probe my body and mind to determine how I have dealt with my experiences as a prisoner of war.

Through their evaluations, based on the clinical diagnosis of Post Traumatic Stress Disorder, psychiatrists have identified three groups of POWs and have concluded that one group considers their imprisonment in Vietnam as an evil event, and goes through life wallowing in despair, hiding their feelings, suppressing and denying that their captivity ever happened.

The second group looks at the event as the central element, the crowning achievement of their life. They thrive on the experience, market it, and ride it to the max making money on speaking tours.

The third group has been identified as having Post Traumatic Stress Growth. These men have kept the event in perspective. They understand that bad things happened to them, grew from it, and went on with their lives.

I believe I'm in that last category.

I got through the ordeal believing that what happened to me was something I couldn't do anything about. I wasn't going sit around feeling sorry for myself or crawl into a shell. I put those long, torturous days in a North Vietnamese prison behind me and never tried to fill in the gap of time lost. I have moved on with my life and done the best I could as a human being, husband, father, and citizen, achieving the rank of a two-star general. And did I mention that I'm still the world's greatest fighter pilot?

Pain

From my experience of imprisonment, I learned that the body is an amazing creation. It can tolerate a lot more abuse than most people realize. It is resilient with the capability to bounce back and knows just how much abuse it can take before its mechanism to cope kicks in.

All of us have childhood stories about a spanking or a trip behind the proverbial woodshed for a bit of "correction" of attitude or a nudge to amend a particular behavior. While a spank on the fanny for one is merely a momentary sting soon forgotten, for another it is a tremendous discomfort that can leave a permanent fear of another such occurrence.

Much has been written about the specific torture techniques we lived through, from simple beatings to hanging suspended from ceiling hooks to being tied into grotesque positions. One was called the Vietnamese Rope Torture where our hands were tied behind our back and our arms rotated upward over our head until they touched our ankles resulting in the arms becoming dislocated. Believe me, that is pain.

But there is another kind of pain that has an even more profound impact. Emotional pain. For years most POWs were not permitted to send or receive mail from their families. Men were left to wonder and worry about loved ones. Fathers missed the birth of children, countless tee ball games, dance recitals, and birthdays. They were forced to endure the agony of knowing their children were growing up without their influence and support. Husbands felt the pain of separation from wives, missed goodnight kisses, and anniversaries. This, too, was pain.

And still, there is another form of pain. Not knowing from day to day if you might be beaten to death, killed by one of your own countrymen dropping a bomb on the prison camp, or having to stand by and watch as fellow prisoners were tortured with no means to help; that's psychological pain and it crawls deep inside where the scars don't show.

Pain, whether physical, emotional, or psychological, leaves a lasting affect. The pain resulting from the torture persists still today in the arthritic joints and broken bones many of us carry in our aging bodies, but the joy of our return home and the camaraderie we enjoy from our common experience largely masks the physical discomfort.

I'm No Hero

I still become uncomfortable when people call me a hero. I don't see myself as being any different than anyone else. I can't leap over tall buildings or bend steel with my bare hands; I'm not superman. I just happened to be one of the lucky guys that God took good care and blessed with a homecoming.

Of the sixty-six pilots in Europe who "volunteered" for a tour of duty in Vietnam, I was the thirty-third shot down. Of those thirty-three, only four of us lived to come home. I was fortunate not to be killed and went on to "enjoy" the POW lifestyle. Not a great job, but it beats the alternative. Those of us who came home in 1973 had the good fortune of not being craters on a hillside in Vietnam. We aren't the ones with our names etched on that granite wall in Washington, D.C.

All of us who were in the direct line of a missile or bullet and ended up as a prisoner of war were ordinary men who had the "opportunity" to experience something we hope no one else will ever have to endure. As Admiral Jeremiah Denton said upon his return, "We are honored to have had the opportunity to serve our country under difficult circumstances."

After hearing about the torture I suffered, the first comment most people make is, "Oh, I could never withstand that kind of pain; I could never survive that kind of torture. Just the thought of my shoulders being twisted and pulled from their socket like a chicken bone . . . no, I could never do that. It's beyond my imagination."

I felt the same way on June 13, 1967. The idea of being beaten with a rubber hose or having a lit cigarette burned into my flesh was beyond my imagination, too. But it happened, and I survived it all beginning June 14, 1967.

When asked how I handled being a POW, I tell folks that each man dealt with his confinement differently. Each of us, individually, and as a group, did whatever it took to survive. We knew we had to get through the ordeal, no matter what, so we could go home to our families. We came up with creative solutions for survival and discovered that we all had physical and emotional strengths far beyond that for which we gave ourselves credit. Mary Head said it best in the book *That's Outside My Boat* by Charlie Jones and Kim Doren, "There's something deep and enduring within all of us that is so strong that unless we have been challenged we'd never have known we have it."

You truly don't know what you're made of until you've been tested. In my case I was fire tested, made into a better human being because of my time in Vietnam. For me, when faced with surviving or dying, dying was not an option. That's not heroic, that's just plain old German stubbornness.

Yes, I spent a total of 2,077 days, give or take a day or two for leap year, in deplorable conditions, but I came home with relatively minor complaints. What about the young men who lost an arm or leg or suffered the effects of a bullet wound in the stomach and are now in hospitals or are seriously handicapped for the rest of their lives? What about the men who didn't have the chance to grow into old age with their loved ones, or see their children and grandchildren grow into adults? They are the real heroes.

So, don't call me a hero.

Random Thoughts

Ernest Hemmingway is considered one of the most competent and versatile users of the English language. With his words he could paint pictures in the reader's mind as vivid as any oil painting. His words aroused feelings so intense they seemed real rather than mere words printed on a page. When asked why he never used that talent to describe the glories, frustrations, and beauty of sex, he explained in effect that if you have been there, you know what it's like and you don't need words to describe it. If you haven't, no words can explain it.

That's how I feel when asked, "What was it like to be a prisoner of war?" I can't describe it. If you were there, you know and don't need words. Each of us had experiences so personal and unique that we find ourselves incapable of adequately relating those experiences and feelings even to our closest friends and family members. Only another POW can understand and appreciate what we went through. And because we can't describe that experience, nobody can understand the emotional bond we have with each other. All we have to do is look each other in the eye and we know and understand what the other one endured. There's no need for words.

Everett Alvarez Jr., the first U.S. pilot shot down and captured, said that we built a cloak around ourselves as a form of protection. To some degree, I still keep an insulated shield around me. It's not that I'm standoffish, it's just that there are some very private matters that I care not to share with anyone else.

Among the lessons learned as a POW was something that I held to as a squadron commander. You achieve greater goals as a team. By yourself you're pretty frail, but when it comes to bouncing back emotionally, it takes a team. That's how we survived. If you can take the "I" and the "Me" out of the equation, and look at the "We" and the "Us," you can survive the most terrible ordeal. *I* didn't survive the horrors of prison; *we* survived.

The men I served with in many ways and many times saved my puny butt. Thanks to the camaraderie, I made it through one day at a time. I left Vietnam with a greater appreciation of the friends I made and served with. I will never forget them and honor their friendship and cherish my time spent with them. I don't take our relationships for granted.

I also developed a deeper regard for the simple things in life. The light switch is on the wall opposite the hinges of the door, four feet from the ground. The hot water faucet is on the left, cold is on the right. I didn't have a light switch in my cell and I didn't have hot or cold running water. Most people in the United States don't appreciate all the comforts they have. But I do, and don't take them for granted.

As Paul Galanti said, "There's no such thing as a bad day when you have a doorknob on the inside of the door."

People need to know that at no time did we leave anyone to suffer alone. Sure, we were physically separated and in solitary much of the time and certainly alone in the interrogation rooms when being questioned or tortured. Still we always had the comfort of knowing others understood, had been there before us, had set a standard of conduct to which we would be held accountable, and ultimately would nurture and comfort us when thrown back into the cell. No one goes through life alone. No one is expected to operate totally as an independent agent. We in uniform understand the concept of team, the role of the wingman and the reality that failure comes soon to those who can't seek the help, companionship, or support of family or friends. Don't ever count yourself out of any situation. You have untapped strength and unseen wingmen.

Throughout many lonely hours spent in solitude, I discovered that the creativity of the human spirit is unbridled. With the richness of what was in my mind when I didn't have the distraction of a television or magazine or newspaper; when I could think and dream freely, I could entertain myself, keep my mind busy, and keep myself sane. Perhaps that's a good lesson for us to teach to our children.

Thanks to the Weasel Club, which forced me to reevaluate my priorities, I came to understand and appreciate what is important in life. When working again, I began to look at life in a different light and see the fun that can be had and not worry about the everyday, mundane occurrences. That's one big lesson learned.

If I had a personal motto it would probably read: Enjoying life is a question of integrity, values, principles and priorities, and keeping them afloat equally on a sea of one's favorite adult beverage.

What a wonderful country we have where people can express their views without reprisal. I didn't come away from living in a Communist society, where the utter repression of freethinking was the norm, to say "different strokes for different folks." Our government may have its faults, but it's still the greatest one in the world as Winston Churchill once said. I didn't have any personal animosity toward anyone who may have walked up to me and called me a warmonger.

In fact, I have addressed the student body at Kent State and was warmly received despite my topic on the war in Vietnam.

I will say, however, that I was ashamed to be hailed a hero while other honorable veterans came home to have rotten eggs and garbage thrown at them. The antiwar protester had no understanding of what it was like to be shot at or how it felt to see your buddy fall next to you with his guts blown away. I have empathy for those young men who went through hell exposed to horrors they weren't emotionally prepared to experience. They were not only traumatized by the savagery of war but also by a hostile society—the very people they were protecting—that branded them baby killers. There wasn't a mechanism in 1973 that identified Post Traumatic Stress Disorder and they got little, if any, help.

That was the real tragedy.

As for the young men who took an antiwar stance and fled to Canada, a lot of them made that decision at a point in their life when they may not have been mature. Right or wrong, it was their decision. Who am I to judge?

One final random thought:

What does Ed Mechenbier live by? I believe one must be true to his values and principles. I like to think that I still live by the principles taught at the Academy.

> *Respect:* Consideration and appreciation for another's beliefs, opinions, and feelings.
>
> *Dignity:* Treating all people in a manner that reflects their worth as a human being.
>
> *Courage:* Courage to do the right thing for the right reasons.

And I always keep to the line from my high school alma mater, "To thine own self be true."

CHAPTER 26

Awards and Honors

I HAVE BEEN GIVEN MANY AWARDS THROUGHOUT THE YEARS, such as the Kentucky Colonel Award; Clement Austria Community Service Award; Ralph Winkle Memorial Award; Honorary Chief Master Sergeant; Bogie Buster Red Jacket recipient; Honorary Citizen of Dallas; Presented the keys to the cities of Cincinnati, Dayton, Xenia, and Springfield.

An award that I am most appreciative of, although I doubt I fully deserve, is the Heritage Award presented in 2000 by the Air Force Association, which is presented annually and given in recognition to the person who worked to promote Wright-Patterson Air Force Base community relations and who was their ardent spokesperson. I truly cherish any and all recognition because it provides me a platform to give a "shout out" to those folks who do the heavy lifting while I receive the attention. I treasure the accolades and plaques, but I'm always aware there are lots of people who know my warts.

I received another recognition during the twentieth annual induction ceremony on November 2011 when I was inducted into the Ohio Veterans Hall of Fame along with nineteen other deserving veterans. Ohio governor John R. Kasich and former Vietnam POW Thomas N. Moe, presented me with a plaque and a red, white, and blue ribbon with a medallion.

But of all the awards I have received over the years, the one I'm most proud of is the Distinguished Graduate Award presented to me by the USAF Academy Association of Graduates in 2006. Of the fifty thousand graduates I was ninth to be selected. A granite pedestal with a plaque mounted on top stands on the Academy grounds highlighting my accomplishments and contributions to the Air Force Academy.

I wear eight rows of ribbons on my uniform beneath my Command Pilot's Wings. They are the Distinguished Service Medal, Silver Star with oak leaf cluster, Distinguished Flying Cross with oak leaf cluster, Bronze Star with "V" device, Purple Heart with oak leaf cluster, Meritorious Service Medal, Air

Medal with a silver oak leaf cluster and three bronze oak leaf clusters, Air Force Commendation Medal, Prisoner of War Medal, Vietnam Service Medal with eight service stars, Republic of Vietnam Gallantry Cross with Palm, and the Republic of Vietnam Campaign Medal.

What Others Say About Ed Mechenbier

"I met Ed in 1973 at Wright-Patterson Air Force Base in Dayton the night he returned from nearly six years as a prisoner of war in North Vietnam. I was a new reporter at WHIO-TV and tried to interview him as he walked into the base hospital that night. I remember that Ed was not well, but he was friendly and polite.

"For almost 30 years, Ed Mechenbier and I spent most of every July working together on the local and national telecasts of the Dayton Air Show. Ed was the one who made the programs work. His knowledge, enthusiasm, and sense of humor were an immediate hit with viewers. "General Ed" was so good that within a few years he was the television voice of many other cities' air shows as well.

"Over the years it became clear to me why Ed Mechenbier wanted to put air shows on television. He genuinely wanted to share the magic of aviation and the accomplishments of the aerospace community with the world—especially with younger Air Force members who didn't have the time or money to bring their families to an air show.

"Ed worked hard in other ways to explain aviation. Away from the air show, he was always there for me and other reporters when we had questions about flying or the military.

"And he's a competitor! During the Dayton Air Show, the "live" TV part of our duties every year would be five or six hours long. It became a matter of honor with each of us not to leave our desks once the program started. So our viewers missed one of the most entertaining parts of the afternoon: Mechenbier and Baldridge scrambling for the porta-potties once the cameras were off!

"Ed suffered horribly in an unpopular war, yet he is free of bitterness. He achieved the highest level of his profession as a general officer in the United States Air Force, yet he is humble. He's a devoted family man and a respected contributor to his community.

"Ed Mechenbier is an authentic American hero."

Jim Baldridge
Retired WHIO-TV News Anchor

"It was very difficult to be positive and encouraging in a North Vietnamese prison camp, but I never heard Ed complain, even once. His leadership in volunteering for the toughest jobs, sense of humor and gentle encouragement of us all made each day more tolerable. He was chosen to be a chaplain because his

Christian Faith was clearly evident in all he did. I believe that every one of the forty-five men in that cell would say the same."

Guy Gruters
Captain, USAF

"I was a television technician for WLW-TV in Dayton, Ohio, the night Ed Mechenbier returned home in what was known as Operation Homecoming, which happened to be the coldest night in Dayton history. The plane was late arriving and we had to scurry around filling live prime airtime until the plane arrived. We began interviewing everyone who would talk to us. One of the reporters interviewed Ed's brother-in-law who warned him that Jerri wasn't one to abide by military protocol. Although she had been told to stay in the car until the official ceremony was over, she didn't wait but bolted from the car and almost knocked Ed over as she ran into his arms.

"A photo of me with a tear frozen on my face as I watched this historic and heart warming event appeared on the front page of the newspaper the next day.

"Later that summer my wife and I had tickets to see the Kenley Players at Memorial Hall in Dayton. We were sitting in the balcony when we noticed people below us standing and clapping. We wondered what was going on. We discovered that Ed and Jerri had just walked in. There was no announcement of any kind, but people had recognized Ed and began paying him the respect he so rightly deserved. I remember thinking that I would love to meet him. He must be one hell of a nice guy.

"Three years later, I was involved with the Dayton Air Show with WHIO channel 7. We were looking for someone who knew about aeronautics and military aircraft, so my boss contacted Ed. I had told several people around the station about how nervous I was to finally meet a real hero.

"Knowing how nervous I was to meet him, the staff told Ed of my trepidation. When he walked into the conference room he took one look at me and said, 'Are you the son of a bitch that took me off the front page?' That's how I officially met Ed Mechenbier."

Bob Romond
WHIO Executive Producer, Dayton Air Show

"To say that Ed Mechenbier is a good golfer is like saying *War and Peace* is just a book. He is devoted to the game of golf and I've never known anyone to work harder at the game. For a guy over sixty and be able to hit a golf ball as straight as he does with a score that matches his age is impressive.

"He has a gentle way of helping fellow golfers if they're having a bad day. He treats people with kindness. He has been like a surrogate dad to me. If I have issues or questions I can always text him and know I will get an answer. He always has his Blackberry smoke'n. If matched with a thirteen-year-old kid, he would give them a run for their money every time."

Rick Webster
Commander, 445th Operational Support Squadron
Wright-Patterson Air Force Base, Dayton, Ohio

"In 1970, just after the U.S. commando raid on the Son Tay POW Camp, the majority of POWs were hastily moved to the old prison camp in Hanoi, Hoa Loa, or as we called it, The Hanoi Hilton. I ended up in cell 6. After a few days we started getting organized, one of the first things we did was start communicating with the other cells to find out how many were in each room, etc. etc. This was late November early December 1970. The cells were laid out in such a way that cell 6 only had a common wall with cell 7, so we could use the tap code between 6 and 7. At the other end of cell 6 there was a gap that was sort of a passageway between cells 4 and 5. The only way to communicate with cell 4 was for someone to stand on a pile of blankets and peer out a high window and look down into cell 4. Each cell had communicators, I was one of them in cell 6 and Ed was one of them in cell 4. I stood there for some time before he noticed me, and then we established communication via the hand code. Ed and I spent many hours looking at each other sending endless messages, Ed was very fast with hand code, but I was slow as hell because I'd never been in a cell where the code could be used, so Ed had to be patient with me until I got up to speed.

"We used to send some long messages from the high rankers at the other end of the camp. The official messages had to be perfect, so the guy receiving the message, (me) had to whisper each word to another POW who wrote the message on a plate. Then the official message had to be sent back to the sender to make sure it was exact. This involved some long stints of communication, sometimes the messages were pure bullshit (in my opinion), and you could tell what the other communicator thought so too by the pained expression on his face. So, that's how Ed and I became acquainted, flashing hand signals to each other and using facial expressions. This sounds slow, but two good hand code senders and receivers could send messages faster than a person could write down what was being said."

John Heber "Spike" Nasmyth, Jr.

"Throughout his imprisonment, he did not lose his fighting spirit. He did not lose his sense of hope. And he did not fail to remain anything but a shining example of a warrior whose duty assignment had been temporarily changed. . . . I'm proud of Ed Mechenbier. I am proud of his accomplishments, but perhaps more importantly, I appreciate his unwavering sense of duty, honor, and country for it is those values that define the warrior spirit. I thank him for the many sacrifices he made for our great Nation . . . "

[Taken from comments made by Senator Mike DeWine of the floor of the United States Senate, June 17, 2004.]

CHAPTER 27

Tribute to Kevin Joseph Patrick McManus

I THANK GOD AND KEVIN EVERYDAY for his courage and clear thinking when our situation went sour over the Vu Chu railroad yard. His persistence in telling me, "Ed, I don't think we're going to make it," saved my life. That is no exaggeration. That's why I consider him a saint.

If I had to spend five years cooped up in a prison cell with anyone, Kevin would be my choice. He was the perfect roommate—for comic relief if nothing else. It's hard to imagine two people, whether husband and wife, or best friends, living in such confined quarters as we did and never, ever having an altercation. There were days that didn't go as smoothly as others. You'd think that somewhere along the way one of us would have angered the other, and even come to blows. But it never happened. Never. We made better use of our time trying to get through each day, holding each other up physically, emotionally, and spiritually.

Kevin was my "rock" for years while we were in our cramped two-man cell. We forged a bond that most people would never understand—one of support and friendship stronger than some husbands and wives. In a normal situation there is always somewhere you can go when things turn bad. Locked in a tiny cell with another guy, you can't exactly escape from each other. As POWs we formed a shroud, a protective shield around ourselves, but my shield embraced Kevin, too.

In 1973 when we said goodbye to each other at Travis Air Force Base, we made a bet as to who would be the first to have a kid. I cheated and adopted, so the bet didn't count. He and Mary Jane went on to have seven children, nine shy of the number they had planned. So, we both bought a bottle of brandy that would go to the first who had a grandchild.

Kevin stayed in the Air Force, achieving the rank of lieutenant colonel. Throughout the years we talked weekly on the phone and saw each other at reunions and, of course, he came to my retirement celebration. After retiring from the Air Force, he became a successful business executive with Robbins-Gioia Company in Washington, D.C., where we often teamed up on projects.

Throughout the years, Kevin's health became an issue—back problems, hip replacement, and heart bypass surgery. So he wasn't in the best of health when he was told he had lung cancer.

I learned that Kevin was seriously ill in the spring of 2008, but didn't know his condition was life threatening until Jerri and I went to see him in Washington that June. He looked quite ill, but characteristically, despite his pain, he still greeted us with a warm smile.

Treatment for the cancer shrunk the tumor temporarily, but it returned. He died in July 31, 2008, and was cremated. The family held a good old-fashioned Irish wake with 150 friends celebrating the life of one hell of a great guy. I gave his eulogy at the Mass held on August 12 at Saint Timothy Catholic Church in Virginia. He was interred with full military honors at Arlington National Cemetery next to his son Rory who died in 1999.

When Kevin died, I already had grandkids but he still had the bottle of brandy. His sons and I broke it open and drank a salute in his honor.

I keep a prayer card for Kevin on the headboard of my bed. In all my speeches, I always include a story about the two of us, along with a couple photographs of us as we left the prison and again as we stood together saluting the officers and flag when we landed at Clark Air Force Base in the Philippines.

Kevin lived by the USAF Core Values: Integrity First, Service before Self, and Excellence in All We Do. He never waivered from his adherence to the Code of Conduct, and when faced with torture and the threat of death, he conducted himself with courage. He always had a quiet air about him that said he knew this world was only temporary and the real life is in the hereafter.

Giving his eulogy was the hardest thing I have ever had to do.

More people need someone like Kevin in their lives—that special individual—who can impact the core of their soul. I miss him terribly.

How do you say goodbye to a dear friend?

2-2, 3-3, 2-2, 1-2, 4-5.

Good Night Kevin. God Bless You.

APPENDIX

Cast of POW Characters

JOSEPH S. ABBOTT, JR., Air Force Major, F-105D pilot assigned to 333rd TFS, 355th TFW at Takhli, Thailand, shot down 30 April 1967.

GARETH LAVERNE ANDERSON, Navy Lieutenant (jg), F-4B radar intercept officer assigned to the USS *Kitty Hawk,* shot down 19 May 1967.

EDWIN LEE ATTERBERRY, Air Force Captain, RF-4C pilot assigned to the 11th TRS, 432nd TRW at Udorn, Thailand, shot down 12 August 1967.

JAMES WILLIAM BAILEY, Navy Lieutenant (jg), F-4B RIO assigned to the USS *Constellation*, shot down 28 June 1967.

WILLIAM JOSEPH BAUGH, Air Force Captain, F-4C aircraft commander assigned to the 480th TFS, 366th TFW at Da Nang, South Vietnam, shot down 21 January 1967.

JAMES ROBERT BERGER, Air Force Major, F-4C pilot-systems operator assigned to the 480th TFS, 366th TFW at Da Nang, South Vietnam, shot down 2 December 1966.

ROBERT IRVING BISS, Air Force Captain, F-4C aircraft commander assigned to the 559th TFS, 12th TFW at Cam Ranh Bay, South Vietnam, shot down 11 November 1966.

JON DAVID BLACK, Air Force Captain, F-4D aircraft commander assigned to the 555th TFS, 8th TFW at Ubon, Thailand, shot down 27 October 1967.

RONALD GLENN BLISS, Air Force First Lieutenant, F-105D pilot assigned to the 357th TFS, 355th TFW at Takhli, Thailand, shot down 4 September 1966.

DONALD RAY BURNS, Air Force Major, F-4C aircraft commander assigned to the 389th TFS, 366th TFW at Da Nang, South Vietnam, shot down 2 December 1966.

MICHAEL THOMAS BURNS, Air Force First Lieutenant, F-4D pilot-systems operator assigned to the 433rd TFS, 8th TFW at Ubon, Thailand, shot down 5 July 1968.

HARLAN PAGE CHAPMAN, Marine Captain, F-8E pilot assigned to the USS *Oriskany*, shot down 5 November 1965.

JOHN WALTER CLARK, Air Force Major, RF-4C aircraft commander assigned to the 11th TRS, 460th TRW at Udorn, Thailand, shot down 12 March 1967.

EARL GLENN COBEIL, Air Force Captain, F-105F pilot assigned to the 357th TFS, 355th TFW at Takhli, Thailand, shot down 5 November 1967.

JAMES QUINCY COLLINS, JR., Air Force Captain, F-105D pilot assigned to the 68th TFS 6441st TFW at Takhli, Thailand, shot down 2 September 1965.

KENNETH WILLIAM CORDIER, Air Force Captain, F-4C pilot-systems operator assigned to the 559th TFS, 12th TFW at Cam Ranh Bay, South Vietnam, shot down 2 December, 1966.

JOSEPH CRECCA, JR., Air Force First Lieutenant, F-4C pilot-systems operator assigned to the 480th TFS, 366th TFW at Da Nang, South Vietnam, shot down 22 November 1966.

JOHN OWEN DAVIES, Air Force First Lieutenant, EB-66C Electronic Warfare Officer assigned to the 41st TRS, 432nd TRW at Takhli, Thailand, shot down 4 February 1967.

JEREMIAH ANDREW DENTON, Navy Commander, A-6A pilot assigned to the USS *Independence,* shot down 18 July 1965.

JOHN ARTHUR DRAMESI, Air Force Captain, F-105D pilot assigned to the 13th TFS, 388th TFW at Korat, Thailand, shot down 2 April 1967.

JEFFREY THOMAS ELLIS, Air Force Captain, F-105D pilot assigned to the 469th TFS, 388th TFW at Korat, Thailand, shot down 17 December 1967.

JOHN HEAPHY FELLOWES, Navy Lieutenant Commander, A-6A aircraft commander assigned to the USS *Constellation,* shot down 27 August 1966.

JOHN FER, Air Force Captain, EB-66C pilot assigned to the 41st TRS, 432nd TRW at Takhli, Thailand, shot down 4 February 1967.

HUBERT KELLY FLESHER, Air Force Major, F-4C aircraft commander assigned to the 480th TFS, 366th TFW at Da Nang, South Vietnam, shot down 2 December 1966.

HENRY POPE FOWLER, JR., Air Force First Lieutenant, F-4C pilot-systems operator assigned to the 433rd TFS, 8th TFW at Ubon, Thailand, shot down 26 March 1967.

LAURENCE VICTOR FRIESE, Marine Captain, A-6A bomb navigator assigned to the VMA (AW)-533, MAG-12 at Chu Lai, South Vietnam, shot down 24 February 1968.

NORMAN CARL GADDIS, Air Force Colonel, F-4C aircraft commander assigned to the 588th TFS, 12th TFW at Cam Ranh Bay, South Vietnam, shot down 12 May 1967.

PAUL EDWARD GALANTI, Navy Lieutenant, A-4C pilot assigned to the USS *Hancock,* shot down 17 June 1966.

DANNY ELLOY GLENN, Navy Lieutenant (jg), A-4C pilot assigned to the USS *Kitty Hawk*, shot down 21 December 1966.

GUY DENNIS GRUTERS, Air Force Captain, F-100F pilot assigned to the 612th TFS, 37th TFW at Phu Cat, South Vietnam, shot down 20 December 1967.

LAWRENCE NICHOLAS GUARINO, Air Force Major, F-105D pilot assigned to the 44th TFS, 18th TFW attached to the 6234th TFW at Korat, Thailand, shot down 14 June 1965.

THOMAS RENWICK HALL, JR., Navy Lieutenant (jg), F-8E pilot assigned to the USS *Bon Homme Richard,* shot down 10 June 1967.

DOUGLAS HEGDAHL, Navy Seaman washed overboard from the USS *Canberra* 6 April 1967.

DONALD LESTER HEILIGER, Air Force Captain, F-105F Wild Weasel navigator assigned to the 13th TFS, 388th TFW at Korat, Thailand, shot down 15 May 1967.

JAY ROGER JENSEN, Air Force Major, F-105F, Wild Weasel navigator assigned to the 13th TFS, 388th TFW at Korat, Thailand, short down 18 February 1967.

JAMES HELMS KASLER, Air Force Major, F-105D pilot assigned to the 354th TFS, 355th TFW at Takhli, Thailand, shot down 8 August 1966.

RICHARD PAUL "POP" KEIRN, Air Force Captain, F-4C aircraft commander assigned to the 354th TFS, 15th TFW at Ubon, Thailand, shot down 24 July 1965.

THEODORE FRANK KOPFMAN, Navy Lieutenant Commander, A-4E pilot assigned to the USS *Ranger,* shot down 15 June 1966.

GALAND DWIGHT KRAMER, Air Force First Lieutenant, F-4C pilot-systems operator assigned to the 390th TFS, 366th TFW at Da Nang, South Vietnam, shot down 19 January 1967.

MICHAEL CHRISTOPHER LANE, Air Force First Lieutenant, F-4C pilot-systems operator assigned to the 559th TFS 12th TFW at Cam Ranh Bay, South Vietnam, shot down 2 December 1966.

GORDON ALBERT "SWEDE" LARSON, Air Force Lieutenant Colonel, F-105D pilot assigned to the 469th TFS, 388th TFW at Korat, Thailand, shot down 5 May 1967.

JAMES D. LATHAM, Air Force Captain, F-4E aircraft commander assigned to the 435th TFS, 8th TFW at Ubon, Thailand, shot down 5 October 1972.

JOSE DAVID LUNA, Air Force Captain, F-105F electronic warfare officer assigned to the 354th TFS, 355th TFW at Takhli, Thailand, shot down 10 March 1967.

JERRY WENDELL MARVEL, Marine Major, A-6A pilot assigned to the VMA (AW)-533, MAG-12 at Chu Lai, South Vietnam, shot down 24 February 1968.

DAVID PAUL MATHENY, Navy Ensign, F-8C pilot assigned to the USS *Oriskany,* shot down 5 October 1967.

KEVIN JOSEPH PATRICK MCMANUS, Air Force First Lieutenant, F-4C pilot-systems operator assigned to the 390th TFS, 366th TFW at Da Nang, South Vietnam, shot down 14 June 1967.

WILLIAM HARLEY MEANS, JR., Air Force Captain, EB-66C pilot assigned to the 41st TRS, 460th TRW attached to the 355th TFW at Takhli, Thailand, shot down 20 July 1966.

READ BLAINE MECLEARY, Navy Lieutenant (jg), A-4E pilot assigned to the USS *Hancock,* shot down 26 May 1967.

JOSEPH EDWARD MILLIGAN, Air Force First Lieutenant, F-4C pilot-systems operator assigned to the 433rd TFS, 8th TFW at Ubon, Thailand, shot down 20 May 1967.

JOHN HEBER NASMYTH, JR., Air Force First Lieutenant, F-4C pilot-systems operator assigned to the 555th TFS, 8th TFW at Ubon, Thailand, shot down 4 September 1966.

NORRIS MILLER OVERLY, Air Force Major, B-57B pilot assigned to the 8th TBS, 405th FW attached to the 35th TFW at Phan Rang, South Vietnam, shot down 10 September 1967.

THOMAS VANCE PARROTT, Air Force Captain, RF-4C navigator assigned to the 11th TRS, 432nd TRW at Udorn, Thailand, shot down 12 August 1967.

JAMES GLENN PIRIE, Navy Lieutenant Commander, A-4E pilot assigned to the USS *Hancock,* shot down 22 June 1967.

JOSEPH CHARLIE PLUMB, Navy Lieutenant (jg), F-4B aircraft commander assigned to the USS *Kitty Hawk,* shot down 19 May 1967.

FREDERICK RAYMOND PURRINGTON, Navy Lieutenant (jg), A-4C pilot assigned to the USS *Franklin D. Roosevelt,* shot down 20 October 1966.

RICHARD RAYMOND RATZLAFF, Navy Lieutenant, F-4B radar intercept officer assigned to the USS *Enterprise,* shot down 20 March 1966.

HERBERT BENJAMIN RINGSDORF, Air Force First Lieutenant, F-4C pilot-systems operator assigned to the 559th TFS, 12th TFW at Cam Ranh Bay, South Vietnam, shot down 11 November 1966.

James Robinson Risner, Air Force Lieutenant Colonel, F-105D pilot assigned to the 67th TFS, 18th TFW attached to 6234th TFW at Korat, Thailand, shot down 16 September 1965.

Wendell Burke Rivers, Navy Lieutenant Commander, A-4E pilot assigned to the USS *Coral Sea,* shot down 10 September 1965.

Mark John Ruhling, Air Force Captain, RF-4C navigator assigned to the 14th TRS, 432nd TRW at Udorn, Thailand, shot down 23 November 1968.

James Richard Shively, Air Force First Lieutenant, F-105D pilot assigned to the 357th TFS, 355th TFW at Takhli, Thailand, shot down 5 May 1967.

Robert Harper Shumaker, Navy Lieutenant Commander, F-8D pilot assigned to the USS *Coral Sea,* shot down 11 February 1965.

Jerry Allen Singleton, Air Force First Lieutenant, CH-3C helicopter crewmember assigned to the 38th ARS at Udorn, Thailand, shot down 6 November 65.

Wayne Ogden Smith, Air Force First Lieutenant, F-4D pilot-weapons-systems operator assigned to the 435th TFS, 8th TFW at Ubon, Thailand, shot down 18 January 1968.

Donald Ray Spoon, Air Force First Lieutenant, F-4C pilot-systems operator assigned to the 480th TFS, 366th TFW at Da Nang, South Vietnam, shot down 21 January 1967.

Charles Davis Stackhouse, Navy Lieutenant, A-4C pilot assigned to the USS *Bon Homme Richard,* shot down 25 April 1967.

Hugh Allen Stafford, Navy Lieutenant Commander, A-4E pilot assigned to the USS *Oriskany,* shot down 31 August 1967.

John Edward Stavast, Air Force Major, RF-4C aircraft commander assigned to the 11th TRS, 432nd TRW at Udorn, Thailand, shot down 17 September 1967.

James Bond Stockdale, Navy Commander, A-4E pilot assigned to the USS *Oriskany,* shot down 9 September 1965.

Hervey Studdie Stockman, Air Force Lieutenant Colonel, F-4C aircraft commander assigned to the 390th TFS, 366th TFW at Da Nang, South Vietnam, shot down 11 June 1967.

Richard Allen Stratton, Navy Lieutenant Commander, A-4E pilot assigned to the USS *Ticonderoga,* shot down 5 January 1967.

Timothy Bernard Sullivan, Navy Lieutenant (jg), F-4B radar intercept officer assigned to the USS *Coral Sea,* shot down 16 November 1967.

Thomas Wrenne Sumpter, Air Force Major, EB-66C navigator-electronic warfare officer assigned to the 41st TEWS, 355th TFW at Takhli, Thailand, shot down 14 January 1968.

ORSON GEORGE SWINDLE, III, Marine Captain, F-8E pilot assigned to the VMF (AW)-235, MAG-11 at Da Nang, South Vietnam, shot down 11 November 1966.

JACK HARVEY TOMES, Air Force Captain, F-105D pilot assigned to the 354th TFS, 355th TFW at Takhli, Thailand, shot down 7 July 1966.

JACK LINWOOD VAN LOAN, Air Force Major, F-4C aircraft commander assigned to the 433rd TFS, 8th TFW at Ubon, Thailand, shot down 20 May 1967.

ROBERT EARL WIDEMAN, Navy Lieutenant (jg), A-4E pilot assigned to the USS *Hancock,* shot down 6 May 1967.

Bibliography

Cross, Coy F., II. "Operation Homecoming: MAC's Finest Hour." Beale AFB, Marysville, Calif., 9th Reconnaissance Wing History Office, February 11, 1999.

Hobson, Chris Michael. *Vietnam Air Losses: United States Air Force, Navy and Marine Corps Fixed-wing Aircraft Losses in Southeast Asia 1961–1973*. England: Midland Publishing, 2001.

Nasmyth, Spike. *2355 Days a POW's Story*. New York: Orion Books, 1991.

Rochester, Stuart I., and Frederick Kiley. *Honor Bound: American Prisoners of War in Southeast Asia, 1961–1973*. Annapolis, Md.: Naval Institute Press, 2007.

Scott, Arcuri A., Major USAF. "Performance of American POWS in The Vietnam War: Adequate Training or Creative Leadership?" Research Report submitted to the faculty, Air Command and Staff College, Air University, Maxwell Air Force Base, Montgomery, Ala., April 2005.

Index

I

J

K

N

T